BECOME A

GAME CHANGER

The BUSINESS PLAYBOOK for the INTERNET ERA

AF874199

MARK BRIGDEN

Become a GAME CHANGER
© Mark Brigden 2015

Disclaimer

The information in this book is not to be taken as a recommendation, but simply as an outline of some strategies, products, services, and websites available in the marketplace at the time of writing. They are provided as examples and you should do your own investigation to determine what suits your circumstance.

All rights reserved. No part of this publication may be reproduced, stored in a retrieval system, or transmitted in any form or by any means electronic, mechanical, photocopying, recording, or otherwise, without the prior written permission of the author.

National Library of Australia Cataloguing-in-Publication entry (pbk)

Author:	Brigden, Mark, author.
Title:	Become a GAME CHANGER : The BUSINESS PLAYBOOK for the INTERNET ERA / Mark Brigden
ISBN:	978099433813 (paperback)
Subjects:	Success in business, Technology
Dewey Number:	658

Published by Mark Brigden and InHouse Publishing
www.inhousepublishing.com.au

Printed using Envirocare paper

Dedication

I won't name all the individuals, however to everyone who has helped me along the way, I greatly appreciate it. To those who have inspired me, know that I am forever in your debt. It is amazing how inspiration comes both from sources and at times you may not expect. Small things can have a big impact, moments can create memories, and ideas become strategies. In addition to this my clients keep me motivated, young, and creative. I love the momentum they exhibit. My family and friends give me the reasons and support to push on. I'm sure you all know the part you've played. I hope you know it's valued.

MARK BRIGDEN

About the Author

I was born and raised in Brisbane and adopted at birth by a loving family who provided a stable upbringing. I'm married to the lovely Monica, with one great daughter, two dogs, and two cats (and, previously, two of the world's slowest racehorses). I've worked in a bank, in lending and marketing state administration roles, headed up a diversified financial services company, and had a great job in international distribution and marketing for an investment fund. Outside of my corporate life I've always had creative outlets, including hosting a weekly radio sports show, creating a football podcast, working as a soccer journalist for newspapers and magazines, and spearheading several online business ventures.

I've travelled extensively, and while there's no place like home, we still feel the need for an annual visit to Hawaii (whenever I can hear the waves calling my name). I like to feel the sun on my skin and be able to see the water.

I live and work in Redland City in South East Queensland (the greatest part of the world), as a business coach and strategist, which gives me the ability to help others achieve their dreams and their targets. I also do a lot of coaching online and want to utilise my membership site to teach others the skills they need to take advantage of the current opportunities out there for them.

I love that I get to work with inspiring business leaders to help them take their business to the next level, by challenging their thinking and helping them adopt new strategies, while they enjoy the process. I help them develop their discipline and ability to envisage, articulate, and achieve their goals. In conjunction with their business success they become leaders in their community and understand the impact they have on others. Seeing their success is mind-blowing to me.

Working with me is a progression. From seminars to training programs available on my *www.brightstrategies.com.au* site, to one-on-one coaching with a select few, and finally the Business Game Changers membership section and webinars, leveraging your talents to generate passive income or raise your credibility.

Whatever your desired destination, I help with the journey. Together we will make a difference.

This book is designed to increase your awareness of your potential, help you overcome any negative mindsets, and show you the stepping stones on the path to success.

Some of you may want some help with this process and that's what we're here for, so contact us for some coaching. Hopefully this book will give you the inspiration, so I'll work with you on the strategy and delivery aspects. Alternatively, if you have a great idea and are looking for someone to partner with on this journey, just let me know. Don't let this stop you from achieving your dream. This is the best opportunity you'll ever have.

Now sit back, relax, enjoy the book, and let your mind roam.

NOTE:

Before you start reading, I strongly suggest that you have a notebook handy. There are some sections near the start of the book where you may wish to write your answers to some questions. Writing these down will help clarify your thoughts and deliver a stronger end result for you. Plus of course you'll want to take notes of the great tips throughout the book, so you may as well keep it all in one place. Grab that notebook and get ready.

Contents

So What is a Game Changer?

1. A Game Changer has new and different ideas that stand out from the crowd and change the outcomes.

2. A visionary strategist who uses creative innovation to alter their business plans, or conceives an entirely new plan by exploring new opportunities and different products.

You can become a Game Changer.

Chapter 1
Your Timing is Perfect

"The two most important days in your life are the day you are born and the day you find out why."

— Mark Twain

I was brought up in a different world—well, not literally, but it seems like it. Back then, you were encouraged to get a job straight out of high school, hopefully with the public service or a bank, because then you had a job for life. My dad, and his dad before him, both spent their entire careers with the Railways Department. Security. Stability. Predictability. University or further studies were only for specialists. In recent years the facade of that security has been exposed, and more businesses are downsizing or looking to outsource various functions.

Don't get me wrong, the corporate world can be a great training ground. You can be ambitious and progress up the ranks, but the best you'll be is comfortable. For the large majority, however, you feel trapped. I know I stayed at the bank for so long because I had access to a concessional home loan interest rate. That half a percent, or whatever it was, meant I was tied to the corporation.

The longer you're working somewhere you don't want to be, the more you lose confidence in your ability to do something else. You become entrenched, your thinking becomes narrower, your drive fades. You lose that childhood aspect of dreaming and believing anything is possible. Your frameworks have closed in and are limiting you.

I long believed the education system in Australia had been designed to churn out good employees, but did little to encourage innovation, creativity, and thinking outside the box. Things have certainly improved in recent years, and the kids of today have access to incredible resources. But there's no reason why everyone can't take advantage of these new technologies.

The Internet and associated technologies have already transformed the way we gather information, communicate, conduct business, and relax. It shapes our perceptions, changes our routines, and has an ever-increasing presence in our lives.

And this is just the start of the transformation. Like any new innovation, technology follows an implementation cycle. The early developers break ground with their discoveries and creation, however they lack the expertise, scale, and resources to fully commercialise it. This takes place some years later when other organisations further develop the use of that technology for commercial applications.

So what am I leading to? This is the crux of my message: now is the time. Now is your time. This is a once-in-a-lifetime opportunity for you to take advantage of. The current situation presents the most favourable circumstances you'll see in your lifetime—not just for a few years, but for your generation. And do it before the younger generation have the life experience to take advantage of the business opportunity the current explosion in technology provides for us.

What am I talking about? I'm talking about being a Game Changer. The chance to do something you love, to make a difference, to follow that passion, or to raise your game. Whatever it is—and we'll do some work on getting clear about it—this is the perfect time for you to step outside the stereotype society has placed on you and feel the freedom to create whatever it is you desire.

One of my favourite business sayings is, "In a gold rush, you want to be the one selling shovels." Instead of speculating, you're selling products in a period of high demand. I can't attribute it, as I've heard it from several sources, so I'm not sure who used it first. However, I love the concept, and it's just as applicable in this new, connected world. Let me give you an example. Everyone now has a mobile phone or a digital camera, and many fancy themselves as a bit of a photo buff, but it's difficult to turn it into a business as there are now a lot of mums trying to become photographers and charge for their services. But the money these days is in information services, so why not run a course teaching others how to use it effectively? You could run a photography course for locals to attend, a guaranteed source of income, rather than trying to sell your photography services. That sounds better, but again, that's trading your time for money.

So, a Game Changer would use technology to leverage their time across a much wider audience, by maybe developing a video of their course and making it available as a training program online. Think that could be successful? Think people would pay for that? Well, just go to one site, *www.udemy.com* and you'll see 6,150 people (as of mid-June 2015) have enrolled there in a photography masterclass costing US$199. That's income of US$1,223,850 for the small amount of time it took to shoot the videos and lay out the training course. This means recurring income for the creator, no more work required, except maybe promoting it when he has the chance. And bear in mind that's just on one website, which I'm sure you and your friends have not even heard of previously.

This by the way has grown quickly—back in March 2015 there were only 2,613 registered students with income at that stage of US$ 520,000. Has your income grown by over US$700,000 in the last three months? That's a Game Changer!

By the way, that's not even close to being the most popular course on that site. In excess of 399,000 people have enrolled in some courses. The beauty of this of course is that projects like this can be completed in your spare time. Once you change your thinking, you can use your experience to create a product (in various formats) and get away from trading your time for money. And this example is just one of many different streams you can utilise, depending on your preferences. Once people have bought a digital product from you, you have their contact details and the ability to develop further products they may be interested in around the same topic.

Current Stresses

Think about the wasted hours in your current job. Think about the impacts on your lifestyle, the travel cost, the commuting time, and the stress. In this book we'll look at options to give you more time with your family, more time doing things you want to, more time being productive.

Even if you decide to stay in employment and dedicate your life to your current firm, you can benefit from utilising these skills and tactics to climb the corporate ladder faster. Many businesses emphasise the way things have always been done and their internal competencies, over an understanding of the evolving needs of the customer. Can you think outside the box? Can you position your business to where technology is heading, without having the head-in-the-sand approach? Can you be more productive by doing part of your job at home, utilising the latest technology? If it is office administration type work, they may allow remote workers; maybe just one day a week to start with, until you can prove the benefits. The smart business of the future will learn to stop measuring inputs (hours worked) and start measuring outputs (achieving outcomes), and there should be no physical boundaries on where that needs to take place.

Or maybe you could consider doing something in conjunction with your day job, to build an income stream. Diverse sources of income are crucial to ensure you don't join the ranks of pensioners like 95% of Australians once they reach retirement age. You have experience. You have a talent. Use that to share your knowledge and generate a passive income stream. Remember, the great majority of people couldn't be bothered to make the effort to do something challenging or different, so you'll already be ahead of them, and this is your chance to escape the world of 'normal'.

Alternatively, you may already be running your own business, but things have plateaued. You might be stuck in a rut working in your business and not taking advantage of the strategic opportunities available these days. It's easy to be overwhelmed with the day-to-day operations, the paperwork, tax compliance, and staff issues. If you're in this situation, who has time for planning? Well, obviously things have to change, or you'll be consumed in a tsunami of smarter strategies. Bright strategies, like the ones I develop for my clients.

Although, here's another way to think about it. Is it really a business, or is it just a job, with yourself as the employer? Try taking three months off and see how it performs without you there. Yep, it's just a job, but you can't slack off, because you're also the boss. In fact, it's much worse because of all the extra responsibilities you have, so no doubt the dream of freedom that your own business would provide got crushed along the way.

There are Options

With technology these days, many services can be provided online, and this can be established with automation in place to do most of the activities without your input. You can be travelling the world, sitting on a beach in Hawaii with your laptop and just checking on progress (and your bank account). That is the new lifestyle business!

I heard a great term the other day—'coffice'. That's when you spend time at a coffee shop using their free Internet access, whilst you soak up the atmosphere and use your laptop to work on your business. This is the life of a traveller, or someone who has established their business so they can still operate it from anywhere in the world. But it's also of great benefit to the solo worker. Mix up your days by spending some time at local coffee shops surrounded by sounds, colourful characters and inspiration. Sounds attractive, doesn't it?

Most people seem to live their life, and their business life, in a state of distress. Does that sound right? Distress is unhealthy and unproductive. Is that you?

Here's another great new term I heard recently: 'eustress'. This is a term coined by an endocrinologist named Hans Selye. The word 'eustress' consists of two parts. The prefix 'eu' derives from the Greek word meaning either 'well' or 'good'. When attached to the word stress, it literally means 'good stress'. So it is used in terms of good stress that motivates you to continue working. And that's exactly what a Game Changer is looking for; work with meaning that they love, and in an environment that they enjoy. Good stress—isn't that the change we are all seeking?

Do you think age has anything to do with reinventing yourself? A State of Entrepreneurship Study by the Kauffman Foundation in 2015 found that the baby boomer generation, those born between 1946 and 1964, are twice as likely to be gearing up to launch a business in the next twelve months than their younger counterparts. While some boomers are still aiming to change the world, many are more focused on factors like independence, pursuing a passion, working less, or supplementing their income. Combine this with their life experience, their love to travel, and their desire to make a contribution, and you have the perfect attributes to start an online venture.

I love the concept of the 'bucket list', made popular in a movie of the same name in 2007. Billionaire Edward Cole (Jack Nicholson) and car mechanic Carter Chambers (Morgan Freeman) are complete strangers, until fate lands them in the same hospital room. The men find they have two things in common: a need to come to terms with who they are and what they have done with their lives, and a desire to complete a list of things they want to see and do before they die. Against their doctor's advice, the men leave the hospital and set out on the adventure of a lifetime.

Indeed, you can check out the dreams of others and get inspired at *www.bucketlist.org.* Maybe you can take advantage of technology to help you work through your own bucket list.

Former World Ironman champion and Australian surf lifesaving legend Guy Leech, believes there are two crucial elements to being successful in anything—understanding your 'why' and letting that drive you; plus utilising whatever legal, unfair advantage you can. In sport, that is an understanding of sport science and the latest training techniques, and in the business world he's talking about technology; utilising things your competitors don't have access to, or haven't yet discovered.

I love this concept of utilising whatever legal, unfair advantage you can. Maybe it's the competitive spirit in me, the side of me that likes to win, or maybe it's the knowledge of something you have over your competition. Think of the America's Cup success

Australia had with the famous winged keel boat way back in 1983. That was our 'huge improvement in technology' at the time, and the mindset around that gave us a huge advantage.

Changes are Coming

Historically, improvements in technology have somehow led to job creation in the past, however a study by researchers at Oxford University predicts that as much as 47% of work that is done today in the United States is at risk of being taken over by computerisation. No doubt that same percentage will apply worldwide, certainly within the developed countries.

Within Australia, a report by Pricewaterhouse Coopers, *Future Proofing Australia's Workforce*, estimates almost half of current Australian professions are at high risk of disappearing, thanks to technology, over the next two decades. That's a lot of jobs—5.1 million to be precise. They believe most of the jobs people are employed in today simply won't exist in the next decade. The positive side though, is that this will allow different opportunities, smaller corporate structures, and new industries to flourish, provided you are aware of the need to transition.

In business, a Game Changer embraces technology. Facebook may be great as a social tool to keep in touch with friends, but social media also provides excellent business opportunities. YouTube, Twitter, Pinterest, and all the other social media options adopted by the younger generations are actually fantastic tools to utilise for business development. LinkedIn is more than just a platform for finding a better job.

Another story I love is that of Jack Smith, who ran a UK-based company called Shyp at a time when they were trying to secure the final allocation of a $100,000 grant from an accelerator program. Aware that thousands of other companies had applied, Jack knew he had to do something different to get the attention of the CEO giving the money away. Consequently, he prepared a LinkedIn advert with a photo of the CEO, and targeted the ad at both that company and six other people he ascertained knew the CEO. The CEO saw the ad, clicked on the link, and saw a special video presentation pleading Shyp's case. Jack Smith got the $100,000, and his innovative advertising method had cost less than $1. Now that's some return on your investment. He went on to repeat it several times to attract the attention of other influential CEOs before LinkedIn tightened their rules.

Whatever path you want to take, now's the time to freshen your thinking and utilise some creativity to take your business to the next level. There's been a fundamental change to the way the economy is run and any high performer can adapt. Now's the time to live your purpose.

So it doesn't matter what level you're at, what age you are, or what industry you're in. What does matter is your mindset. Are you ready to be a Game Changer?

Mark My Words

(Let's recap and pay attention to these important parts)

- Don't allow your limiting frameworks to stop the success you deserve.
- Technology has transformed the way we gather information, communicate, conduct business, and relax.
- Your knowledge and experience are invaluable tools.
- Use the leverage of technology to engage mass markets.
- Create the lifestyle business you want.
- Use any legal, unfair advantage you can in business.
- Adapt to the fundamental changes brought on by technology.
- The timing is perfect for your new adventure.
- You now have the opportunity to do something different.

Chapter 2
The Simplest Formulas Can Be Magic

"Don't limit yourself. Many people limit themselves to what they think they can do. You can go as far as your mind lets you. What you believe, remember, you can achieve."

—Mary Kay Ash

J. Paul Getty had a formula for success: "Rise early, work hard, strike oil."

If only we could; but unfortunately that's not going to pay off for most of us. Maybe these days you could replace the striking oil part with winning the lotto. 'Rise early, work hard, win the lotto'. Of course, the odds of that happening are over eight-million to one, so the statisticians tell us you have more chance of dating a supermodel than winning the lotto, and I wouldn't advise putting your hopes in either basket.

(Dreams are okay, but don't make unattainable ones your plans.)

Wouldn't it be great if there was a magic formula? I mean, we all want the easy way out, don't we? Let me tell you, I think there is an easy way out. It's simply this:

Success is a question of choice and behaviour.

When you think about it, this makes sense. It's not that we're all unintelligent. For example we all know that to lose weight we need a combination of regular exercise and good nutrition. We know it, but how many of us actually act on it?

It comes down to making the *choice* to proceed and then altering our *behaviours* to meet our goals. Those that do change their behaviours reap the benefits with some great results, while the rest of society wants to blame others, e.g. 'healthy food is too expensive'.

Even when you start something, such as your New Year's resolution, it's hard to maintain that intensity for a period of time. Seriously, how many of you last even to the end of January? And the reason for that is that motivation is short-lived. Like those memes with great sayings that bombard your Facebook account, a burst of motivation might brighten your moment, but it doesn't sustain change. You need inspiration to change.

Until you take action, nothing changes. But as humans, we're programmed in a complicated way. The brain is unbelievably powerful, but we limit its use. So you have to smash through the mental roadblocks that sabotage your future. You can master new behaviours that will have a huge impact, and you will see it's not as difficult as you think. Don't fall into the habit of compounding ignorance—you don't know and you don't know that you don't know. If you want it enough to learn some personal development skills, make sacrifices, be disciplined, and understand we operate in a new paradigm, you will find a way to be successful.

There are some basic steps along the way, including:

- Know what you want
- Know why you have to have it
- Commit to a great strategy
- Measure your progress daily
- Keep doing it until you reach your goal

Take these steps and add two core skills crucial to delivering you success—influence and leverage.

By influence, I mean the ability to position yourself as an expert and deliver a compelling message. And leverage is the way we utilise technology to reach a much wider audience than previously possible.

Now what do we do to adapt this to the new technological space? I mean, it all seems so complicated, doesn't it—who knows where to start? Well, sometimes we just overlook the obvious. The quickest way to be successful is to model someone, or a business, that's already hit the heights. Why reinvent the wheel? They have used a model, a process that has been proven successful, so you just need to mimic/replicate it—making appropriate changes, of course. We'll be more specific about this in later chapters, but for now you can relax in the knowledge that there's a format for you to follow.

Overcoming Your Boundaries

Before you can overcome the frameworks that may be limiting you, an understanding of your drivers is appropriate.

Let's first determine what it is you want:

1. Most will say **money**, and in many respects they're right. Money might not make you happy, as the often repeated claim goes, but what it does is give you the freedom to choose. And let's face it, who couldn't do with more money and the lifestyle options that brings? You can then become the master of your own destiny. Unfortunately, being employed or the old ways of doing business are not going to deliver the income you want. Of course there will always be successful businesses, but they still need to adapt to compete these days. So if money is your driving force, then becoming a Game Changer is a no-brainer.
2. When you get a bit older you realise just how precious **time** is. Time to do the

things you want, to fit it all in to your busy schedule. Is this possible in your current work situation? Then the answer is using the Internet, or associated technology, to change your delivery mechanism and leverage the ability to reach a mass audience. Then with automation you can create time; time for yourself.

3. Others crave **self-worth**. A feeling that their life means something; that their skills are recognised and they have made a contribution. Albert Einstein recognised this when he said, "Strive not to be a success, but rather to be of value." This is the perfect scenario for you to become a Game Changer. Use your skills and life experience to influence and benefit others. Tell your story, share your knowledge; it will provide value to your audience.

The stumbling block around money, for most people has been their upbringing. Lessons learnt at an early age manifest themselves in your psyche, so the attitude you saw in your parents and the way they approached money and business will have had a bigger impact than you realise.

The other strong influence is your peer group. For most people, they can calculate the average income of their five closest friends, and that is similar to what their own income will be. We simply push ourselves to the limit of those we surround ourselves with. Try it, prove me wrong if you can. But despite the influence of your family and friends you can push through these limiting factors. Seek out those that inspire you, those that have been successful or have already achieved the challenges you set yourself. Their approach, their belief systems, and their attitudes are contagious. This way you're listening to voices of value, to people who have the experience, ideas, and success you're striving to emulate.

My upbringing has enabled me to clearly see the difference between negative beliefs and empowering or positive thoughts. My mother had a very rough childhood and suffered from nervous problems. As such I grew accustomed to hearing negative beliefs, conservative concerns, and worrying aspects. She is an amazing woman who, at eighty-five, has managed to live on her own after Dad died twenty years ago, and cope with the problems of maintaining a house. Yet she only sees the negative side of things. Of course her thought processes are dictated by a lifetime of challenges, but it clearly showed me, growing up, that there were two ways to think about every issue. And it's amazing how many negative aspects revolve around money, which stem from childhood issues that stuck with her forever and restricted her enjoyment of life.

If I do nothing else in this book than encourage you to broaden your outlook and understand the difference between acting like a victim and realising you have control over your life, then I'll still be very happy.

I've seen the difference it can make to people. Don't let your restricting beliefs stop you from having the future you deserve.

I completely understand that emotional issues affect your mindset, and the effect your environment can have on you growing up. But you have to realise that clinging to those beliefs is what's standing in the way of you being the best that you can be. Your thoughts dictate your life circumstances. You become what you think about. It's as if you are programmed to respond to certain situations in a particular way with a predetermined mindset.

Those who take the victim, defeatist, or 'I'm just unlucky' mindset tend to act in a manner consistent with those beliefs, and thus stop themselves from being successful. These things tend to become a self-fulfilling, self-perpetuating prophecy.

It's important to understand you can only control three things:

- The thoughts you think
- The pictures you see
- The actions you take

Change your daydreams, your habits, your beliefs. This includes the conversations you have in your head (your self-talk), which affects your self-belief, your body language, and the stories you tell, which impacts on how others perceive you. Can you see how it flows on?

You either create or allow things to happen to you, so changing your thought process manifests this. Just take responsibility for making some changes and you actually have the power to make it happen.

Think of the mental preparation that goes into being a sports champion. If they make a mistake they are taught not to focus on it, to keep their attention on the end result, and to always look forward. They are taught to envision their success and the steps along the way. It's the same in business and life. Don't dwell on past mistakes. Don't let your self-talk become self-doubt.

This is basic psychology, but a key message to understand is that *your thoughts influence your decisions, which create your actions, which then drive your results.* So, if you change the input at the start—your thought processes—you dramatically change the rest of the equation.

You can try doing this with some subliminal messaging software, which flashes messages you have created onto your computer screen to reprogram your subconscious mind. This is a good way of programming your intentions and thus allowing your subconscious mind to continually reaffirm these positive thoughts. For example, you could set it up to flash messages such as 'I will finish my book within the next month', or 'I will earn $10,000 per month from my online venture', or even, by the time you finish this book, 'I will be a Game Changer'.

This is certainly a powerful method of influencing your mind to look at positive thoughts. I've used it myself. Even if you don't agree with the concept, you can set it up to flash at such a quick speed that you don't recognise it with your conscious mind. Then you can still claim the success as your own.

Another option is to be selective with the content that goes into your mind. By that I mean the books you read, the films you watch, and the people you pay attention to. Make sure your role models are positive people. It's better to admire Oprah and Dr Phil than most footballers or musicians. I realise that's just a generalisation to make a point, and you can always be surprised. For example I used to love the group Kiss, and they were one of my favourite bands behind The Angels and ACDC—yes, I'm showing my age! Think about the band members and reflect on what your impression of Gene Simmons used to be. Yes, the one with the tongue. Well, you wouldn't know it, but Gene is a smart marketer and a hugely successful businessman across a range of industries. His book *Me, Inc.* is a great story of how to win in life and business. So yes, you can always be surprised, but be as selective as you can about the people and media that influence your thinking.

Simple Approach

Let's go through a process in a manner that enables you to understand it.

It's important to first establish why you want to make money. Really, what can you do with it? What changes will it allow? Try this simple exercise, and you need to write your answers down as that helps clarify your thoughts, rather than just reflect on it. The fact you have to write your answers down also forces you to think harder about this aspect, which is exactly what you need to do.

List twenty-five reasons why you **must** make more money.

Once you do this exercise, and force yourself to list at least twenty-five reasons—and I'm sure you can come up with a lot more if you focus—it should open your eyes.

It becomes pretty clear now that the old ways haven't been working and that this is your opportunity to finally achieve something special. Something has been holding you back and stopping you from making the money you need.

Now what would you use that extra money for? What can you see yourself doing? What's the dream lifestyle you want to lead every day? Grab that pen again.

Describe the daily routine you aspire to. Some things might be monthly, or even annual, but give some thought to the range of things you want to experience and how you'd lead your life.

Basically, once you get clear on this, you've completed the first two steps:

1. Know what you want.
2. Know why you have to have it.

In many cases the reason you haven't taken any bold steps previously is because of fear:

- Fear of failure
- Fear of the unknown
- Fear of rejection
- Fear of stepping outside your comfort zone

But now, your focus is on outcomes and presumably what you can do for others. That's powerful. That trumps fear. That gives you the inner strength to change.

Let us summarise that and write it down.

I want to be a Game Changer because ...

Before we start to get creative about what you'd like to do, let's just reinforce those dreams you've had along the way. We've all been constrained for too long and that certainly inhibits our dreams. I realise that many of the things you wish for change over time, depending on the circumstances in your life. And many of the experiences we would have liked to enjoy, we haven't, because we were worried about the perceptions of others. Yes, another great fear that holds us back is the fear of what other people might think. The funny part of this is that in most instances we don't even know these 'other' people, so we are letting people who have no input into our life impact on how we act. It's time you started to do the things *you* want.

I'm sure there are still some things tugging at your heartstrings. Did you wish you had the confidence to sing in public, or travel to an exotic country on your own, or take the plunge into that business idea that's been quietly nagging at you? Be honest with yourself.

Once you've given it further thought, complete this simple exercise.

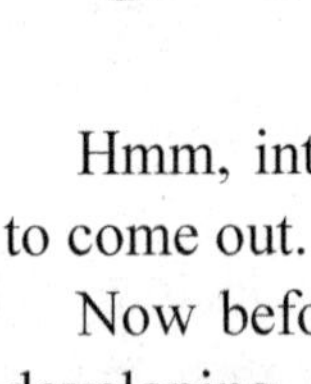

Write down ten things you'd do if you KNEW you couldn't fail, or you wouldn't be judged.

Hmm, interesting. I bet there's still more waiting to come out.

Now before we move on to the exciting steps of developing a great strategy, then building your business and continuing your progress, we need to address how changes can help you develop the groundwork for your future success.

Mark My Words

- Success is a question of choice and behaviour.
- Your success will come with influence and leverage.
- Determine your drivers.
- Don't be afraid to live your dreams.
- Look for positive people and influences.
- Overcome your stumbling blocks by focusing on outcomes or positives.
- Thoughts—decisions—actions—results.
- Spend time writing down your reasons.

Chapter 3
Positioning Yourself for Success

"Some people want it to happen, some wish it would happen, others make it happen."
—Michael Jordan

Like with anything, you have to lay a solid foundation. This includes how you approach the day. Get up early, write down the goals you want to achieve that day, and do a workout. Having a healthy breakfast is crucial. Why? Because you want to perform at your maximum all day and that means you want energy.

You'll notice the difference in the energy levels between successful people and the rest. They look after their fitness and their health because they are more focused, and have a set routine every day. Some of them are also well aware of how and when their body functions best, and consider that when scheduling important tasks. We all have a well-defined internal clock that shapes our energy levels throughout the day: our circadian process, which is often referred to as a circadian rhythm because it tends to be very regular. An example of this at work would be when you're suffering from jetlag, as your body expects certain things at prescribed times. This natural, ingrained rise and fall in energy levels has important implications for when you are required to perform at your best.

Having a set routine in the morning positions you for success, as that is the period the majority of people are most productive. You work on yourself and you work 'on' your business, rather than in it. Thus you're better prepared, more confident, and ready to be more productive. It gives you the chance to decide the important tasks, thus you're being proactive rather than reactive. And isn't that one of the main problems during the day? "We'd all achieve much more if the customers would just stop ringing." Seriously, I hear those complaints often.

Often the reason a business has failed is not because they didn't achieve their goals, but because they weren't clear about exactly what their goals were. This is particularly true for solo operators and small business owners, where the focus is so much more on just dealing with the day-to-day problems. As individuals we like to have direction, meaning, and a purpose—put on your business hat and it should be no different.

A study at Virginia Tech University reportedly found that people with goals earn nine times more over their lifetime than people without goals. Now that's a good reason to give this area your attention! I'm sure though that there are some sceptics out there, so let's be over conservative and say you might only get twice as much income in this scenario. It still makes great sense. And if it is what successful people do, then again, we should model their behaviour.

There are lots of processes available for how you should set your goals. One simple method is to do it in three parts—short term goals, medium term goals, and … no, not long term goals, as things change too quickly in the online world. The third part is what is often referred to as a BHAG, which is a big, hairy, audacious goal. This is where you shoot for the stars and aim for something that would be considered an outrageous success.

From the big picture, to the now. We've all heard about time management. Actually, time can't be managed, but how you use that time can be. So it's actually self-management that you need—now that is something you can control. I'm sure we all wonder how certain people achieve what they do. Can you imagine the workload of Richard Branson, Oprah Winfrey, or even Eddie McGuire? They run networks and various industries, and forever seem available to talk to the media. No, they're not cloned, and no, they don't have more than twenty-four hours a day. They just apply great self-management to their time. It's a discipline and, yes, it's a very difficult skill to master. They make getting to the heart of things the big issue, and they do that by being focused on the key aspect.

The more quickly you can narrow the focus, stop all the distractions, and take action, the more quickly you'll get to the root cause or point of maximum impact.

Eliminate the Padding

Don't confuse being busy with being productive. Those who've had a long stint in corporate life tend to know how easy it is to fill in your day with tasks, without actually achieving anything. Instead of controlling our activities, we find they control us. We don't want that as a Game Changer so we have to eliminate the padding. In fact we want to achieve much more, whilst doing much less. That means you have to set priorities after actually thinking about what you're trying to achieve.

This is similar in many respects to the Pareto Principle, or the 80/20 rule, which basically states that for many events 80% of the result comes from 20% of the effort. (Or, as commonly interpreted, most of your sales come from a smaller percentage of your clients.) So determine what your 20% is and focus on that part. Be the best you can be at that aspect.

Many famous entrepreneurs ask themselves every day whether this particular activity they're considering is really going to assist in driving the company to where they want it to go. If they go through a process like that to question the validity of each action, then you should too.

That's why, first, you need to be very clear on what you're trying to achieve, and understand what the end game is that you want. This gives you the determination to make the most of every day. So let's think about it. How can you structure your day to maximum effect? What are your time wasters? Yes, you need some relaxation time, but how can you ensure your focus is on productive matters in your working hours?

You're always hearing about the importance of making a To Do list, and I am certainly going to endorse that. That should be routine on a daily basis. However, here's a new twist. Your day is full of interruptions and distractions abound—so make a Don't Do list.

What will you eliminate from your daily activities that is not productive for generating the type of success you crave? Can you stop watching the news, or reading the paper? Can you stop checking your emails every five minutes? Can you cut back on Facebook usage, or at least limit it to, say, fifteen minutes in the evening? Let's be honest, we take on a lot of crap that is not necessary for producing the results we need. You've got to make the tough decision. Don't watch TV to relax, do a workout—it'll be more beneficial for you. What else can you eliminate?

Let's make your Don't Do list.

So, what can you eliminate? What personal distractions? Which business distractions? How much extra time per week will that generate?

How will you use that time? Here's a hint—that's time you can allocate to your Game Changer adventure.

Changing Habits

We spoke earlier about the need to do things differently. Here's an example. I love to read at night whilst lying in bed, it's something I've always done. It'll normally be books about either the cover-up behind the JFK assassination, sport (okay, mainly Manchester United), or some creative business books. Now, as you can imagine I'm passionate about these topics. I get engrossed in what I read and it stimulates my mind, so what happens? That's right, I can't get to sleep. I toss and turn, thus I'm not always at my most productive the next day.

So the solution was quite easy. Instead of reading at night, I started reading my books for thirty minutes as soon as I woke up. I slept better, woke earlier, and then was stimulated, feeling creative and positive as I started the day. The flow of energy is very important to your success, so don't overlook the simple things.

Simple changes can have rewarding effects. Remember, earlier I mentioned the easiest way to succeed at anything was to model someone who had already done it.

Many successful people don't waste time reading newspapers or watching much TV. Why? Because the standard of reporting has deteriorated and it's now about ratings and sales, rather than quality information. There is very little that will influence your personal life and even less that will benefit your business. Everything is sensationalised, so you're not getting much information of benefit. Instead successful people look at free apps, such as Flipboard, which provides a digital easy-to-read magazine in which you have preselected the topics of interest to you. So, for example, if you want business news and the latest in technology, that's the content that will be collated in thumbnail form, and you can select what you want to read in a bit more depth. There's actually over 34,000 different topics available, so nearly every niche is catered for, and you can focus on pertinent issues. Thus you're not wasting time reading unnecessary garbage in the newspaper.

Remember, time is like money in the new economy, so use it productively.

Mindset is a daily thing. You have to be in the right frame of mind for every decision or action you take. Many people get hung up on the past and let that impact upon their decision-making process. Just focus on the now. Decide on issues by assessing which option leads to a better future. Which choice is going to get you, and the business, where you want to be?

Here's a great way to help your decision-making process. Whatever the issue, just run it through these three steps:

1. What's the best possible outcome if you choose to take this course of action? Then decide if that's what you really want.
2. What's the worst case scenario of this course of action? How will that option make you feel?
3. Between the two options, is the risk/reward ratio something that you're comfortable with?

This will help you be more decisive, and that's a great attribute to have as many people waste too much time dithering and stressing over options. Great leaders are very good at decision making. Sure, they're smart and experienced, but they also have a mindset that allows them to believe that at that time, whichever option they choose, it is the right decision. Thus they don't second guess, or take a half-hearted approach. They proceed with confidence because of their belief system. Be decisive.

Another powerful tool is attentive listening. Listen to your clients. Listen to those who aren't your clients. Most of us are too busy talking or doing to give our full attention to what is really happening around us. The information you glean, whether it's positive or negative, is of huge benefit to you in helping you make future decisions. What is the market telling you? Why do they feel that way? Don't presume you know—listen to their words.

One way to create momentum is to set a date. That makes it real. Set a date that you have to have your project finished by, a date that puts a timeframe around your work so that now you have to get the ball rolling. Most people waste too much time trying to perfect every aspect of their idea, and suffer what the experts call 'paralysis by analysis'. Whenever I feel the need to undertake a major project, such as competing in a triathlon or writing this book, I tell people about my plans and when I'll have it done by. I make it public knowledge. That way I'm committed to acting by a certain date, which means I have to train, prepare, or start writing, or else I'm going to look foolish (and my competitive instincts won't allow that).

Cementing Habits

It can take time to integrate these new changes into your lifestyle. You have to try to master them before you move on, or else you can slip back into old habits. So just be aware as you do things and always question whether they're in your best interest. Be intentional with your habits, particularly what you study, read, or listen to. You'll learn new skills, but just make sure you're focusing on ones that are congruent with your short- and long-term goals.

Until recently, scientists thought that the brain structure was hardwired and immutable. But new research has shown that the brain is able to change through experiences, as the neuroplasticity is able to change its structure and function. So there's no excuse, anyone can learn and stretch their neural pathways. Each pathway is associated with a particular action or behaviour, and our experiences, thoughts, and actions strengthen these pathways. This then allows habits to form, through repetition, as the brain will find these things easy to do.

A report by Duke University over ten years ago found that more than forty per cent of the actions people performed each day weren't actual decisions, but habits.

Habits are formed through a process called 'context-dependent repetition'. As an example of this, imagine that you decide every morning you want to be healthy and have a piece of fruit before breakfast. When you first eat the fruit after getting up, a mental link is formed between the context (getting out of bed) and your response to that context (eating some fruit). Each time you subsequently eat in response to getting out of bed, this link strengthens, to the point that getting up prompts you to eat some fruit automatically, without giving it much prior thought; a habit has formed.

Habits are mentally efficient. The automation of frequent behaviours allows us to conserve the mental resources that we would otherwise use to monitor and control these behaviours. They are likely to persist over time because they are automatic and so do not rely on conscious thought, memory, or willpower. You can't extinguish a bad habit, you can only change it.

The often-quoted time period of twenty-one days to learn a new habit is a myth. The time varies dependent upon the complexity of the task and your strength of character and determination, but as long as you persist, then a habit will form. And if you are instilling the discipline of good, productive behaviours, the habits that form will help lay the foundation for your success.

Many marketers actually use the psychology behind habits in their campaigns. First they look for a simple and obvious cue, and secondly they clearly define the rewards. Charles Duhigg, in his excellent book *The Power of Habit*, gives some examples of the cue/routine/reward process. If you've been in any shopping mall in America (although they're now also in thirty countries) you'll know the fast food store Cinnabon, famous for their gooey cinnamon rolls. Their stores are always strategically placed away from the food court, as they want the sweet smell to spread widely and thus stimulate your subconscious. Once your brain starts craving it, you can't wait to taste it.

Another instance was sports related, where NFL coach Tony Dungy knew that if he could install the right habits his team would win. He limited the number of plays they would use, thus simplifying their approach, and practiced these continuously until the players' reactions were automatic. He understood the psychology that to change a habit you must keep the old cue, and deliver the old reward, but insert a new routine. Thus their actions become an automatic response to situations following the habits they'd learned.

You might think this is easier with footballers as they just have to respond to situations. However, these responses have to be automatic; they don't have time to think about them, they just have to react immediately. Time is crucial in that game, where split seconds can make all the difference. Now doing the right thing at the right time takes practice—lots of practice—to ensure you're switched on.

Charles Duhigg summed it up when he said:

> We can choose our habits once we know how. Everything we know about habits, from neurologists studying amnesiacs and organizational experts remaking companies, is that any of them can be changed, if you understand how they function. Hundreds of habits influence our days—they guide how we get dressed in the morning, talk to our kids, and fall asleep at night; they impact what we eat for lunch, how we do business and whether we exercise or have a beer after work. Each of them has a different cue and offers a different reward. Some are simple and others are complex, drawing upon emotional triggers and offering subtle neuro-chemical prizes. But every habit, no matter its complexity, is malleable. The most addicted alcoholics can become sober. The most dysfunctional companies can transform themselves. A high school dropout can become a successful

> manager. However to modify a habit, you must *decide* to change it. You must consciously accept the hard work of identifying the cues and rewards that drive the habits routines, and find alternatives. You must know you have control and be self-conscious enough to use it.

I love his words, "You must know you have control and be self-conscious enough to use it." This simple sentence is the foundation for you to change your habits and shape your destiny. You have control. YOU have control. Powerful stuff. The way you think and act will decide your future. You just have to want it enough to change. You can make small steps to start with, pathways to your bigger goal. Using small goals tricks your brain. Your subconscious likes to be in control and doesn't like change; a huge change will usually result in subconscious resistance, whereas a small change will be accepted.

Cravings are what drive habits, so you need to clearly anticipate the reward—only when your brain starts expecting the reward and craving the endorphins or sense of accomplishment, will it become automatic to take the necessary action. Always try to internalise the right thought process.

The key thing to remember, of course, is that until you take action, nothing changes.

> *"99% of all failures come from people who have a habit of making excuses."*
>
> —George Washington Carver
>
> Former US slave turned botanist and inventor

Personal Branding

In this new world, personal branding is more important than ever. It's so easy, and common, to check up on people you might want to do business with. One thing you must do is Google yourself—not just your business, if you have one at the moment, but also your personal name. Statistics say 92% of prospective customers will do a name and business search before doing business with you. So make sure you know what comes up in a search and take action to ensure you present the right image. Does it link to the right webpage? Does it show articles about you? Does it link to videos of you (being professional, not drunk at the Christmas party)? If not, develop the type of image that you want to portray and ensure that content is ranked high in a Google search. Check on Google images in case someone else has tagged you in a photo. Delete any pictures or videos that aren't appropriate. Remove references to any topics that may be considered controversial, including in your social media posts. If you can't delete these things, just make sure there's an avalanche of positive things, so no-one will track back far enough to see any negatives. Ensure your profile is complete and professional and conveys the impression of someone they'd want to do business with. Now you're looking good.

The second thing you do is set up a Google Alert. It's easy, free, and provides excellent intelligence (www.googlealertstool.com). It allows you to set up alerts based on keywords, which can be emailed to you daily. Most people use these for topics of interest, basically to get updates on products, news, and competitors, but you should also have one for your name so you are aware of any online mentions. Forewarned is forearmed. Actually, there's also a free iPhone app you can use, called Hooks, where you can receive alerts about almost anything you're interested in. At this stage most people are just using these for things like an alert about when their favourite television show is on, but they'll soon realise the commercial potential for their own business.

Now there's an app that even takes this a step further. It's called IFTTT (short for 'If This Then That') and is a workflow system hosted on the cloud—this just means your data and applications are hosted on the web, rather than on your computer. This app is utilised where a situation or event triggers an automated response, for example, 'If I post a photo on Instagram or Facebook, then automatically save the photo to Dropbox', or 'If a comment is posted on my blog, forward it to X'. It's all about staying aware and protecting your brand.

While we're talking about personal branding, here's an important point. Have you secured your domain in your personal name yet? As in www.markbrigden.com, or www.jennyjones.com.au? This is important online real estate and for a nominal cost, even if you don't intend doing anything with it at this stage, I strongly recommend you snap it up before it's gone. Make sure that it is registered in your name if someone else is helping develop your website. If you haven't bought a domain name before just check out www.netregistry.com.au, www.name.com, or www.godaddy.com and you'll see the easy steps to follow.

Most of you will already have a LinkedIn profile and a Facebook page, however the great majority of them are amateurish. They're a vital part of your online brand, so you need them to look good. How do you do this without spending money? Simple, just find a profile and layout of someone who is very successful in your industry, but not a direct competitor (they might be based overseas). Then you 'model' it, which just means you want to give yours a similar feel, with a good layout and your personalised content while using them as inspiration. So you don't copy it, but you mimic the appearance and content of theirs. Remember, if people check your profile you want to give them the best possible impression.

Getting extraordinary results comes from creating a domino effect in your life. You've no doubt heard the phrase 'setting yourself up for success' which is what successful people do. They find what works for them to help push them towards their ultimate goal, and repeat that behaviour. They then look for the next thing that will give them an advantage and add it into the mix.

Some of the things we cover later in this book might seem at first a bit technical and hard to understand. But now that you're firm on the need for some changes in what you're doing, you realise the need to push yourself in the technical sphere. Don't fret, we'll show you how to outsource the difficult stuff to others who are specialised in this field.

To help you on this journey, Paul Timms, the Founder and CEO of the National Coaching Institute, emphasises that there are twelve key points that apply not only to business, but any aspect of life.

1. Anything is possible.
2. The future is predictable.
3. Know what matters to you.
4. Get clear about your reality.
5. Create your daily habits and do them.
6. Get a great coach (who believes in you).
7. Focus on actions, not results.
8. Measure your progress monthly.
9. Set bold, specific, and exciting goals.
10. Treat it like a game and play to win.
11. When you start winning, speed up.
12. You can transform yourself.

Game Changers use their intelligence, ideas, relationships, and implementation through taking advantage of technology, to be successful.

With this in mind let's do some work to get you in the right mindset for success.

Mark My Words

- Modelling what successful people do requires discipline.
- Eliminate non-productive activities.
- Make decisive decisions.
- Eliminate bad decisions made out of habit.
- Understand how good habits lay the foundation for success.
- Work on context-dependent repetition to create good habits.
- Know you have control and be self-conscious enough to use it.
- Your online personal image is vitally important.
- Buy your personal name domain.
- Until you take action, nothing changes.
- Momentum builds momentum.

Chapter 4
Get Clear and Know It

"I continue to be drawn to clarity and simplicity, 'less is more' remains my mantra."
—Stephane Rolland
French fashion designer

To demonstrate the importance of clarity and mindset I first wanted to bring in an expert, Natasha Howie, an Identity Specialist. I posed two simple questions to Natasha. I began with:

Mark Brigden: Do people who are successful in business have a different mindset?

Natasha Howie: I didn't take a keen interest in the power of the mind until the age of twenty-two years old, when I discovered firsthand through a traumatic situation that mindset alone could literally break (and later I discovered, make) a human being. What I experienced catapulted me into a consuming passion. I was driven to understand how the human mind works, and how we utilise our minds to shape and create our own lives. This encompasses how we influence the lives of those we love around us, and how we use our minds to impact on our staff, co-workers, clients, and the wider community.

The mindset of successful businesspeople particularly intrigues me, because it is so different (actually in stark contrast) from that of the general populous.

By far the most fundamental difference in the mindset of a successful businessperson is in their sense of who they are and how they see themselves. Their very identity, or self-image, is solidly forged in a way that will always ultimately lead to success no matter what.

Let me explain why this is such an important and fundamental distinction—your identity, or sense of who you are, sits at the very core of your beliefs about yourself and your beliefs about life. This consequently has a dramatic effect on your capabilities, actions, and ultimately your results. It all stems from your identity.

Successful business owners have a very strong sense of self, and a solid understanding of who they are (their identity). They see themselves as the master of their own destiny, with the power to create their vision and the ability to achieve whatever they set out to achieve. They believe in themselves.

One of the most extraordinary examples of strength of identity that I have seen in a successful businessperson is that of Maha Sinnathamby, the visionary who developed Australia's largest master planned city, Greater Springfield, in Queensland. Greater Springfield has also been recognised internationally as the World's Best Master Planned Community 2010.

Chairman Sinnathamby had to overcome what would seem to most people like insurmountable obstacles to create his incredible vision, including purchasing the land without money, orchestrating change in federal parliamentary laws, and dealing with huge setbacks and hardships all along the way.

When he told me his story of how at one stage in his life he was forty million dollars in the red, with no idea how he was going to get through that hurdle, I was compelled to ask the question: 'When you were at your lowest point, what did you say to yourself?'

Maha's humble response was so stunning that I can still hear his words of wisdom resonating in my mind and soul loudly and clearly. 'Oh Natasha,' he said with a beaming smile. 'I am very kind to my best friend. My best friend is there with me as soon as I wake up in the morning until I go to sleep at night, so I am very kind to my best friend. Do you know who my best friend is?'

Well, there it was on a silver platter. The secret of success straight from the mouth of a highly accomplished billionaire socialpreneur. Simple, yet profound and powerful words of wisdom. How kind are you to your best friend?

You see, success can simply come down to knowing that you are a powerful creator and believing that you will achieve whatever you set out to do. This knowing is inherent to our true nature, and we are born with this wisdom inside each and every one of us. Well that's easy then isn't it? Everybody can be successful now. Ah, what complex creatures we are, human beings. Unfortunately, as we grow up and take in our surroundings, we hear messages from our external world to the contrary. The ages of zero to seven in a child's development is known as their formative or imprint years, and is a critical period in the development of our identity. During this time when the negative messages start to outweigh the positive, we start from a very young age to make erroneous decisions about ourselves. This leads to the plethora of labels that end up creating huge roadblocks in our lives.

Imagine if, as children, we had constant reinforcement that we are inherently powerful with infinite resources, and capable of achieving anything we can conceive of? What would that do to your sense of identity? What affect would that have on your results?

Take a few moments to write down the top three 'labels' that you give yourself that hold yourself back. E.g. 'I'm not smart enough'.

Now write down the opposite in the most positive and empowering way you can think of. E.g. 'My mind is more powerful than I could ever imagine, and I'm smart enough to do anything I choose'.

Then write down something that you would do if you removed the disempowering label, again creating the most empowering and positive statement you can. E.g. 'I'm more than smart enough to reinvent my business or start in a new direction, and have lots of fun with it'.

I then moved on to my second question for Natasha.

> **Mark:** I can see the importance of self-talk and self-belief, but how necessary is clarity of purpose?
>
> **Natasha:** That's a really great question. Clarity of purpose is necessary for two very important reasons … 1) Clarity of purpose harnesses the power of our mind, and 2) Clarity of purpose unleashes the power of our heart.
>
> When our mind is clear on our purpose, it engages the power of our reticular activating system, or RAS for short. Most people don't know about the RAS, and it is one of the most consciously under-utilised parts of our brain. The RAS is a network of nerve fibers that originates in the medulla oblongata (brain stem area) and proceeds up through the midbrain into the neocortex, the place where thought is processed.
>
> Put simply, the RAS functions as a filter of information, which allows us to focus. For example, if we are looking to purchase a specific type of vehicle, we start to notice that

particular vehicle pop up on our 'radar' much more frequently than we did previously—we begin to see that particular car more on the road, notice it more in advertisements, and hear about it in conversations.

When we are crystal clear on our purpose, the RAS draws the necessary ideas and information that will help us create that purpose to our attention. It will stimulate us to think about people that we should contact, strategies we should engage, resources we need, and more. Being clear on our purpose gives our minds the framework to start with, then the RAS will build upon that framework.

Once we have the framework for the structure or purpose we want to fulfill, then we need fuel to get it moving! That's where the power of the heart comes in. Having clarity of purpose engages the heart. It makes us feel good about what we are doing. If we are purposefully doing what we love, it liberates a powerful energy within ourselves that fuels passion and persistence, and allows us to tap into our unlimited inner resources. It allows us to express our innate gifts and talents to a greater level. It gives us courage, strength, and resilience we never knew we had. Tapping into the power of our heart through clarity of purpose gives us a sense of achievement and a knowing of who we are and what we are truly capable of.

We can liken having clarity of purpose to creating a high performance car. The mind and RAS creates the structure of the fine-tuned vehicle, and the heart is the fuel that powers the car.

The biggest problem I find with business owners who want to take their businesses to the next level is that they are lacking in clarity of purpose. This lack of clarity within themselves then infiltrates into every part of their business. Their team lacks clarity, therefore lacking productivity and effectiveness. Their marketing will lack clarity as they won't be clear on what is going on in the minds of their ideal customers, causing the marketplace to lack clarity on their brand identity, which definitely reflects on their bottom line.

Clarity of purpose will dramatically turn this around. It causes a powerful alignment between management, staff, suppliers, and customers, and unlocks the massive potential of the business.

Take a few moments to evaluate your clarity of purpose:

On a scale of one through ten, where one is cloudy and ten is crystal clear, how would you rate your own clarity of purpose for your business?

If you didn't get a ten, what would it take to get there?

What simple action steps could you take in the next week to get more clarity of purpose? E.g. brainstorm with a friend, engage a coach, go to a café with the purpose of doing a brainstorming session, ask people questions, write a list of what you love or what's important to you.

Now that we understand the need to reassess our beliefs and get some clarity about purpose, let's start working through that process. Before we do, let me share with you a quote from Sara Blakely. "My dad encouraged us to fail. Growing up, he would ask us what we had failed at that week. If we didn't have something he would be disappointed. It changed my mindset at an early age that failure is not the outcome; failure is not trying. Don't be afraid to fail," she said. Sara is, of course, a hugely successful businesswoman and a billionaire. She founded Spanx, an undergarment manufacturing company, in her late 20s. As you can interpret, her upbringing gave her the courage to try, the support to tread her own path, and the mindset that she wouldn't be defeated.

Wouldn't it be great if we all had that approach to life and to business?

As you grow into this project you'll need to have a growth mindset, which is a belief that your basic qualities are things you can cultivate through your efforts. You may not know the answers yet, but you have the ability to work your way through the issues.

Spiritual Aspects

One way to get clarity about the things that mean something to you is meditation. Now while I'm not personally into most aspects of spirituality, I can see many people benefit from it. In fact it appears that mindfulness and meditation have a great impact on the ability of the brain to create the neural pathways that I mentioned earlier.

I know that the latest research indicates that meditation not only reduces stress levels, but it also leads to dramatic improvements of certain sections of the brain that enable you to focus and learn better. Unfortunately, my mind is always buzzing with ideas and slowing it down is not something I enjoy. In fact I'm often criticised for spending time in my head instead of being fully present with those around me.

Actually, I will tell you about a day of calmness. It was my first trip to Hawaii and I'd spent the night drinking excessively and making new friends. The next morning my mate and I were due to go on a bus trip around Oahu. I put up with it as long as I could, but eventually the travel sickness (that's what I'm claiming it was) got the best of me and I had to get off. Having absolutely no idea where I was, I walked down to a lovely, quiet bay, far away from the tourist strip. I sat under a palm tree and could see some young local kids kicking a football around. Due to my condition I was in no hurry to leave, and ended up spending hours just relaxing and taking it all in. Palm trees swaying in the gentle breeze, gorgeous blue water, stunning sunshine, white sand, and nowhere to be; it couldn't have been any better. That was my Zen moment as I realised all the things that were important to me. The vision of that day is still implanted in my mind over thirty years later. Powerful stuff.

So when you have your quiet moments, you have to use them in ways that will benefit you and help get your thoughts in order. Slow down the chatter inside your head and try to determine the real priorities or key components in any venture.

To be successful, you need to be clear about a few things:

- Yourself, as in your strengths, your talents, and what you bring to the table
- Your business strategy
- What products you will offer

Think of the successful people you know. I bet they seem confident, knowledgeable, and focused. The reason they get what they want out of life is because they know what they want. They're clear on their direction and offering. They're able to hit their goals because they know what they are. The clarity gives them something to focus in on, to make their thoughts and actions laser-sharp.

Before you start any journey you have to get some clarity on what it is that YOU really want. The best way to do this is the now/where/how concept. That is, take a realistic look at your present situation, including your current business or your interests, strengths, and abilities. Then project your mind to imagine how you want to lead your life, what your business would look like, and how all the factors would interact. Until you know exactly where you want to be, and how you want to operate, you can't plan how to get there. You tend to be reactive, rather than proactive. So swap it around and create the life you want from the outset.

To start with it may be hard to imagine what it is you really want, because your mindset has become constrained by corporate shackles. Look at it this way. If you won $10 million in the lotto, what would your lifestyle be like? I'm sure you'd celebrate and maybe travel, but then how would you want to spend your time? No doubt helping others, or doing something that you're passionate about. So what is it that tickles your fancy? Now how can you turn that into a product?

The other way to look at it is to consider your experience, your knowledge, and your skills. Can you use that to help others? We all tend to underrate our attributes, so have a good hard look at what areas you have excelled in over your life.

Now here's an important point. Don't think that you have to develop some earth-shattering concept, as most offerings are just a repackaging of things that have been offered before. It's your take on something, or maybe a mixture of other ideas you've seen, just concentrated in one place. So maybe you can just improve on an old product or service, or even make that information more targeted to a particular niche.

There are some great by-products for yourself as a result of gaining clarity. The main one is that your confidence follows. You know what you want, you know how to get it, and you become more focused. Thus, you are far more likely to succeed if you have clarity. Whilst you're thinking about that, let's add some stimulus to the creative process.

Mark My Words

- Successful people have a positive mindset.
- Eliminate the negative messages.
- Control your self-talk.
- Clarity harnesses the power of the mind.
- Clarity unleashes the power of the heart.
- Your confidence will increase as a result.
- Consider meditation and mindfulness.
- Look to personalise your offerings into a product.

Chapter 5
Spark Your Creative Genius

"Creativity is a great motivator because it makes people interested in what they are doing. Creativity gives hope that there can be a worthwhile idea. Creativity gives the possibility of some sort of achievement to everyone. Creativity makes life more fun and more interesting."

—Edward de Bono

Remember when you were younger you were full of ideas. Of course many of them were probably outside the realm of your control, but surprisingly someone delivered on them years later. Or you were a bit older and you had a great concept in mind, but lacked the confidence and knowledge to take it further. How many people have said, "I remember thinking about that, I knew it would work …"? Well, now's your chance. What's your idea, what's your message, what's your gift to the world?

Good ideas have a blend of empathy and imagination, which means you should have an understanding of what customers really want and look to deliver that in a creative way. It doesn't have to be complex; the simple things are often the most effective.

We all respond to different stimuli to help get us into a creative zone. A positive mindset, or certainly the absence of negative or judgemental influences, is a great start. You should certainly allow yourself some time away from your normal distractions, where you can relax and allow your thoughts the opportunity to roam. Try to look at things from a different angle, a different approach.

Many successful people also keep notes in a special journal for things that they may be able to use in the future. It's a great habit to get into. I certainly do this, usually after I've been inspired by an industry expert and it has set my mind off on a tangent.

Getting creative can be as simple as changing your perspective on things. What to most people is just 'dirt', in the right hands it becomes 'soil than can nourish life'.

Paul Timms uses a simple exercise to get you thinking about possibilities. He suggests there are seven ways you can make a million dollars this year:

1. Sell something that costs $1 and sell 1,000,000 of them.
2. Sell something that costs $10 and sell 100,000 of them.

3. Sell something that costs $100 and sell 10,000 of them.
4. Sell something that costs $1,000 and sell 1,000 of them.
5. Sell something that costs $10,000 and sell 100 of them.
6. Sell something that costs $100,000 and sell 10 of them.
7. Sell something that costs $1,000,000 and sell 1 of them.

What can you envisage within each category? For example, can you develop a simple app that sells for $1?

Maybe it's more realistic for you to sell 1,000 products—that's only, say, eighty-four customers a month, or break it down further to only about 2.8 customers per day. Hmmm, what could you offer to interest them? What niche could you attack? What need could you satisfy? If you're passionate about something I'm sure many others are too. Are they passionate enough to pay? That is the question you have to decide/answer.

Here's another simple exercise—think of it this way. Let's stick with the million dollars theme and look at it from a different perspective. What if you outlaid, say, $1,000 to start an Internet business? Could you turn this into $36,250 in your first year of this new business concept? That's not much of a stretch. Then, as you refine your marketing, you double your income each year. Once you're creating $36,250 in online income a year it's not much of a stretch to get to $72,500. You're getting more exposure via social media and it is just steamrolling. Thus it continues doubling every year as you explore new boundaries, which seems possible. So in just seven years, what would your annual income be? Yes, $1,000,000.

Don't get me wrong, I'm not one of those marketers saying this is your passport to riches. I'm simply pointing out that the power of the Internet is the leverage or exposure it can provide for your business, and the lifestyle changes possible as an effect of this. The Internet allows you to create and distribute products at a miniscule cost compared to in the retail world. It allows you to reach potential clients, those interested in the same niche, in different states and countries. It allows you to run a business without the cost of bricks and mortar premises with overheads and staff, yet provides the flexibility in enabling you to operate it from anywhere. It also allows you to create a saleable asset, not just a job. And maybe, just maybe, you can hit those figures that will give you financial freedom.

Targeting

Now you can see the potential, keep thinking about which markets you may want to serve. You can simply find other products to sell, but it's best to start with one of your own. Most people in information marketing start with an eBook. Remember, all you need to know is more than someone else who desires that knowledge, and they will pay for it. Let's get the excuses out of the way—yes, someone may have already done something similar, but they haven't done it with your perspective on the matter. No matter what you choose, let's say cookbooks, there's already a million of them available, but people still buy new ones. So it doesn't matter what it is, just personalise it.

Start by targeting a well-defined niche, as the more you market your offering, the more it will appeal to their specific interests. Why do the experts say you should focus on a particular niche? It's simple really, and it comes down to how you attract business. If you target fifty-year-old males at that stage of their life where they want to do something meaningful before they retire, your adverts will be written in such a way that they will think you are talking only to them. "Bingo, that's me," they'll think. Compare this to an advert that targets people who want to do something different—it's just wishy-washy in comparison and certainly doesn't pique anyone's interest.

Remember, the advantage in this is the leverage that social media provides us. We now reach a much bigger audience so we can be selective in our target market. If you had a local retail shop and only catered to sixty-year-old women who are left handed, you are extremely limited in your clientele. I'm sure you'd fail. However over the Internet you are opening up a huge market, even with those restricting criteria.

Here's something else that's smart. Your products can be adapted for different target markets. For example a book titled *How to Sell Real Estate* would not have the same impact as *How to Sell Your Property in Brisbane*. It appears much more specialised, more relevant, and thus more likely to be of interest to potential sellers, real estate agents, and others in Brisbane. Then you replicate that product, with some minor amendments, and rename it, *How to Sell Your Property in Melbourne*. That's a simplified example, but you can see the benefits of utilising a niche for higher impact. And of course if you're a real estate agent in one of those areas and you provide a copy of your book, that you wrote, to anyone considering selling and deciding on an agent, how impressive would that be? Much better than handing them a business card. Instantly you appear as an industry leader. You'd be certain to win their business. You'd be someone they want to deal with.

Are you any closer to deciding your niche?

Jennie Gorman, a Mastermind leader and guiding light for small businesses, believes the best way to realise your dream is through discovering your true passion. Her key question is always, "Why are you here and what is your purpose?" Think about that for a moment. Do you know your answer?

I also love one of her sayings, "Know your why and it will make you cry." That certainly seems to be true of many people these days as they seem to be searching for what they're meant to do for their own internal satisfaction. Remember those exercises we did back in Chapter 2 around what you could use money for, what your daily routine would look like, and what you'd do if you knew you wouldn't fail? They will help you determine your 'why'.

Jennie uses her questions to help you determine your vision, which must be clear and concise for it to be manifested exactly as you see it in your thoughts. It is all about being authentic and thus your real self. By understanding your real traits she believes you can find your niche market and transfer your passion to your business. Her book *Awesome Authenticity* examines this in more detail.

Using her logic, there's no way I would, or could, create some material around, say, home renovations. It certainly wouldn't be authentic, nor is it something I am passionate about. I can just imagine me trying to demonstrate or tell someone how to take old wallpaper off a wall with a steamer and sounding credible. It might be fun trying, but I'd soon be found out to be totally inept. I'm the sort of guy who goes to Bunnings just for the sausage sizzle.

What do you do, believe in, or offer that would interest others? This is where your authenticity comes in. Be yourself and you'll attract people who see value in what you say or your beliefs.

Another great example of being authentic came from one of my clients, who is a business coach herself. She was working with a local hairdresser who was, shall we say, way out. You know the type I mean—purple hair, tattoos, all sorts of metal objects in her face. This look is not exactly going to appeal to classy, middle-aged ladies, who you would imagine are the main target market for most hairdressers (particularly as they can pay more). Of course, I'm not being disparaging of this girl's hairdressing talent. So my client, Dawn from Parker-Jones Boutique Salon Coaching, encouraged her to play up her image further. She encouraged the hairdresser to use that as her niche and attract every edgy person out there. The hairdresser wanted to be herself, she wanted to be authentic, so Dawn just utilised that aspect in her marketing. There's nothing wrong with being yourself, so use your uniqueness as your strong point and that gives you a point of difference.

It's also worth noting that people like to be around others who are comfortable with themselves. Be confident, understand what makes you different, and don't go into your shell because you feel different or somehow inferior to others. Your self-esteem is an important tool in your networking.

Branching Out

It may be that time in life when you're ready for your own business. You want to break free, be your own boss, make the right decisions, and do something meaningful to you, instead of the same old mundane stuff every day. So your next thought is, what do I do? One of the popular options previously was to buy a franchise. Okay, which system suits me? How far away is the available site? What are all the ongoing fees? This solution might suit some people, but you're taking on a lot of risk, a substantial ongoing commitment, and not giving yourself much flexibility as you have to follow their system.

Another option has been to go into partnership with someone else to ensure you can afford the upfront costs and have some help with the workload as you build a business. Unfortunately, many experienced people will tell you that partnerships can be fraught with danger, particularly if they involve family or friends. How can you be sure the other person will have the same commitment to the business as you will long term? Do you both have the same exit strategy? Do you have complementary skills that will benefit the business?

Of course, you could always buy an existing business, but then you have concerns about whether the financial statements are correct and whether they can be replicated under your control. Are you buying someone else's problem? Is that the real reason they're selling?

An easier, and more satisfying option, is to tread your own path. We all have dreams. (One of mine is to buy back Vegemite so it's not owned by overseas interests anymore. That's just sacrilege. My other dream involves re-inventing the comic book into a particular niche, but making it relevant in the modern day).

Is it time to do something you've always wanted to? Can you turn this passion into a business? A clear benefit of this of course would be that you will be more driven in your day-to-day activities and it would result in a happier and more engaged life if you make a success of it. You'll feel like you're living and not just existing.

First off, let's look at what a passion is. In this case it's something that generates a powerful and compelling emotion within you. It's the thing that makes you light up when you talk about it. It's a feeling stronger than even coffee (for most people—Swiss chocolate for me though), something you become addicted to, look forward to, and love to experience. Sometimes others see it in your eyes, your voice, and your enthusiasm when talking about the subject, before you even realise it yourself. So if you ever find yourself sitting on a park bench talking for hours about a subject, you can be sure passion is involved.

Let's look at an example of turning a passion into a business. I was at lunch recently for my sister's birthday and was in deep discussion about this with her partner. He has worked for many years in the local council in an administrative role, dealing with plumbers. Thus his recent experience is very narrow in that field and he is only a couple of years away from retiring, but still young at heart and looking for something to stimulate his mind (and life).

So the discussion initially centred around his passions, as he's now looking to do something that he finds personally fulfilling. It was easier in this case because I know him, however we quickly agreed one of his biggest passions was the two-week holiday he took every year to head out bush.

He knows some people who have a rural property of over a hundred thousand acres, which is relatively small for the Australian outback, and miles from any major city. For years he and a group of friends have spent as much time as they can out there, at what they call their 'Bush Hilton', getting away from the demands of modern life. They live rougher, they do manual work to help out, they might sleep under the stars (although they do have old shearers' quarters to camp in), and they eat much simpler food. There are lots of beers, lots of stories, and just good old-fashioned camaraderie. It's his escape from the pressures and modern-day expectations we put on ourselves.

At numerous parties and functions I've seen others enraptured during his stories about what they get up to, and so many others have stated they'd love to go next time. So there's his passion, and some brief market research indicates others are also interested; now how does he commercialise it?

Obviously we would normally spend some time with him to do some strategic planning around this, and I don't want to give away his thoughts, so we'll keep it general. He could start by registering a business name and getting a website domain name that gives a clue to what it's about, say, www.menescapes.com.au, setting up a WordPress site, and getting a graphic designer to help with the logo and look of the site. Then he would start writing some articles discussing the experience and turning his website into a blog. Videos would be a great addition to help set the atmosphere. He could offer a free gift that he would give away in exchange for someone's contact details. This could be in the form of, say, a list of 'the top 10 escape holidays just for men', just to stimulate some desire. Alternatively people could sign up for a free newsletter to get their regular dose of escapism. He would also set up a business page on Facebook (in the same name as the website) and he could also look at other forms of social media, including Google Plus.

He would use social media every day to promote the website and include photos and other inspirational messages. Once he has a legion of followers and people getting his newsletters he can then start to commercialise the idea. He may have organised other properties out west who would like help with general duties and be prepared to house a group. He may have products available for sale, where he just acts as the middleman and takes a cut on sales. He may offer places on the next trip at a price. He may start a cookbook utilising the basic tools available on their outback adventures, or a survival fun guide.

The beauty of this is that he can get this started now, whilst he's still working. It won't cost much at all. It's fun and stimulating, as it is his passion, so it hardly seems like work. He's communicating with people who have a similar interest, and this tribe will just increase his enthusiasm for the project. He certainly won't need motivation, as he probably does on a Monday morning now. He'll be champing at the bit to share the experiences. Plus, he can be authentic. He can have a beer in his hand while he's shooting videos under the bright sky and barren plains. He can dress as scruffily as he likes, as that will appeal to others looking for a similar experience.

Remember, energy is the key to producing great output, and I'm sure he'll receive positive interest and energy back from those interested. Sounds like a win-win situation while he's winding down the corporate clock towards retirement.

What About You?

With all this in mind to help you decide on your topic of interest, answer these five questions:

- What are you passionate about?
- What do you have experience in, or what can you do and understand easily that others can't?
- Who needs to hear your message?
- What's the best format to get it to them?
- What would you include in that package?

Don't make the mistake of undervaluing your life experience. You may have been a housewife for a long time, but you have plenty of skills to pass on to the younger generation. Remember, schooling is different these days, and parents are often too busy to teach life skills or old-fashioned skills to their offspring. So children are not taught how to do domestic duties like baking, sewing, knitting, or even old school tasks like taking shorthand, which means the younger generation may be keen for lessons on topics like these.

Steve Jobs had a useful spin on this.

> Creativity is just connecting things. When you ask creative people how they did something, they feel a little guilty because they didn't really do it, they just saw something. It seemed obvious to them after a while. That's because they were able to connect experiences they've had and synthesise new things.

So use your experiences. You'd be surprised about how many people actually value the things you've done and become proficient at. Alternatively, you might like the concept reported in Hawaii Business magazine and developed by the Institute of Design at Stanford University, called Design Thinking, as a methodology for arriving at creative solutions and out-of-the-box ideas. They use five steps when seeking innovation:

1. Empathy—gather information through interviews, observation, or first-hand experience. Keep an open mind and just listen.
2. Define—narrow it down and describe the specific problem you are trying to solve.

3. Ideate (Conceive)—working within a time limit, hold a blue sky brainstorming session. No idea is too improbable to write down.
4. Prototype—using whatever materials are at hand, quickly throw together models of your top ideas.
5. Test—take your prototypes to the end users and listen to their honest feedback, then repeat steps 4 and 5.

Whether you decide to go with your passion, or utilise your experience, finding out what people want is also much easier these days. Remember, most people buy for pleasure or to solve a problem. Pleasure and pain. You want to create a product that addresses one of these issues, and the more specific you can be, the better your chance of success.

Once you have a general topic in mind, do a simple keyword search on Google (www.googleeasysearch.com) and you will see what Internet searchers look for online. If they're searching for information about your topic, they're seeking answers to a problem. Perfect for you.

If you're having trouble with the transition, think of it this way: what business are you in, or if you're an employee, what business does your company do? Let me give you a hint—every business is the same in this respect, and we're all in this one industry. Yes, that's right, you're currently in the business of marketing. Without marketing there would be no customers for the products you currently provide. Now admittedly at the moment you may have a role in administration or the factory line, but your company is just like every other; a marketing institution trying to sell a product or service, irrespective of whether they have one client or many.

Thus your new Game Changer business idea is no different, just that the product you choose is more targeted to a select niche. It will probably be much simpler, but you will utilise leverage through the Internet, and with some connections we'll discuss later, to greatly extend your reach.

A lot of experts will tell you the money to be made these days is in the digital publishing business, because you can reach many more people than previously possible, at such a low cost. The same can and will apply to many other products and services.

Successful people have found a way to undertake projects that they don't consider work. It's something they enjoy, so they will complete the work aspect (be that writing a book, producing a television show, setting up a new business) and treat it like a game that they love. Thus it doesn't seem like work because they're enjoying it, but then they don't spend years just working in it. They manage the process and oversee the final result. That way it stays fresh and is not all-consuming on an ongoing basis.

And don't think this concept doesn't apply to all industries. Let's look at an example of adding technology to your strategic approach: now, your passion might be clothes and you might have a retail business, so your initial reaction is that digital publishing doesn't apply to you. You might also think your Internet presence can only extend so far, as clients will still need to come into your shop. Obviously you can develop your online presence and utilise social media to promote it more, but you can also look to tweak your business model. You could also create an extra, appointment-only environment, whereby you provide a more personalised service. This could extend to some personal branding talks, backed up with wine and cheese to turn it into a memorable event. So you're not telling them to buy clothes, just what is pertinent about the fashion that suits them. Of course you could video some of this talk, which could be used to raise your business profile via clips on Facebook, YouTube, etc., which could lead to greater things.

Then you could take advantage of a QR (Quick Response) code to create an innovative marketing campaign. The increased usage of mobile phones has enabled smart marketers, in Japan and the US particularly, to lead potential shoppers to either a web address or a YouTube video. This way they are more engaged and you are able to position yourself as the expert. You could produce an advertisement aimed at the fashion-savvy woman and just have the QR code, which should pique their curiosity. Or go the extra mile and go to the extent of personalising every outfit you sell with a QR code on the tag, explaining its features and what other items it would look best with (an online example of 'Would you like chips with that?', just much more professional). This way they pick up an item, and use their phone to see a video discussing its benefits, along with visual content of how great it looks. Then they buy the companion piece, and you've doubled your sales.

In addition to this, you post lots of videos online, including testimonials or women talking about how a certain piece looks—don't try to sell the items yourself, let others do that for you. That will be all the influence the general public needs to buy your clothes online.

You can easily make your own QR code (for free) at www.qrstuff.com or you can become a subscriber and get some great add-ons, such as scan analytics. If you're looking for something a bit more stylish, there are sites that can incorporate the codes into something like your logo. In fact, up to 30% of a code can be distorted or purposefully left out and it will still be readable by most scanning devices.

By the way, the use of Quick Response codes dropped off for a few years, but Snapchat recently introduced a feature called Snapcodes, which has suddenly made QR codes cool again.

Let's look at another example. Gaming over the Internet is popular amongst teenagers and young adults. Two guys who loved gaming were looking at ways they could make money out of it, so what they now do is buy a version of any new game that's likely to be popular as soon as it's released. They then dedicate the next twenty-four to forty-eight hours playing that game until they master it and have learnt the tricks. Next, they film themselves playing, including tips on shortcuts or how to get to the next level, which are available for subscribers on YouTube. Other gamers love these tips and are prepared to pay money to see the videos. The result is over $500,000 profit per year. We used to tease kids who spent long hours playing games that they'd be losers, but we can see who the winners are when they think like a Game Changer.

How about a hypothetical case, if you're interested in sports? You might decide to start a virtual store selling football jerseys and items for fans, without wanting to incur any of the overheads associated with a retail outlet. How could you reach them in a creative way? What if you decide to have exposure at places where people pass in large numbers, such as train stations? You could place large posters in the walkways that would allow people to scan on their mobile phone and have their order mailed to them that day. Of course, once you have developed the technology, you can then put the posters up in walkways at football stadiums, CBD hallways, or anywhere else people congregate.

The quick lesson from these examples, is that there's always a way to lift your game, to be innovative, and to provide a service in a new way.

Remember, profit doesn't come from the passion, it comes from externalising it (taking action).

Your Turn

Now let's get back to your project. To help you stimulate your creativity, once you come up with some ideas, apply these eight questions to them and you will see which ones are more likely to be successful:

1. Does this product (or maybe even a service) solve a problem? Look for ways to explain the different aspects, approach it from different angles and give some well-rounded thought to the users point of view.
2. How can you show that it has been successful? The obvious way is through testimonials, but you could use generic quotes from industry experts about the need for this type of product, or for some products, before and after photos and examples would be beneficial.
3. Can you tell a story about how it evolved and what it does? Never underestimate the power of a story. Your ability to communicate and engage will impact the success of a product or service. Understand the key drivers for your potential clients and use emotion to get to it.
4. What aspects will you include in the product or service? This is where you start to drill down. Say you can identify four areas that your product addresses, then take it a step further and include some bullet points under each category.

5. Can you elaborate on the benefits of using this product or service? Remember, benefits are more important than features in the eyes of a potential client.
6. How does it compare to your competitors? It's important to understand what your competitors are offering, where your product might be different and what you could promote as positive factors in this regard.
7. Can you spin off any other products or generate additional income from it? This is new age thinking, whereby you want to build a relationship with someone interested in a particular niche and then continue to supply relevant products to them. One simple way to do this is to compile a list of questions people might submit to you after buying your product, and that gives you the basis for a follow-up information product.
8. Can you create a good marketing hook for it? As you'll see later, you'll use social media to promote your offering, so a catchy headline that secures their interest is an important aspect.

If it passes those tests you might have a good offering on your hands.

Write the components down, or if you favour visuals, draw a diagram and sketch it out. I'm sure you've heard of a mind map, a diagram used to visually organise information. You start with a single concept and then parts branch off that. Obviously you can just use a blank piece of paper, or you can utilise technology and download some free software from www.xmind.net, which allows you to keep an electronic record of your project ideas and topics. This way you can easily share that document with others if you want to sound out their opinion.

Alternatively you can use a free app called Office Lens, which enables your mobile phone to act like a scanner. Use it to take pictures of notes, receipts, business cards, menus, whiteboards, or sticky notes—then let Office Lens crop, enhance, and save to OneNote. Just like that, all the scanned images you capture from Office Lens are accessible on all your devices.

It's hard to break the habit of writing things down, so another great option coming to the market is Rocketbook, and this is crazy. It's just like a regular notebook, however when you've written on every page you are able to digitise all your notes and store them in the cloud, and then clear all the pages so you can write on them again.

How can this work? Well, the pages of the notebook (which are just normal paper) feature a set of seven different symbols. These symbols can be mapped to different functions, such as saving to Dropbox, Evernote, Google Docs, or even just sending the document to your email inbox. Mark one of the symbols on the page, and when you scan it with Rocketbook's accompanying smartphone app, a digitised copy of the page will instantly be sent to the cloud storage platform of your choice. After you've scanned, digitised, and saved your notes, you can just pop Rocketbook in the microwave for thirty seconds and it'll erase all your notes—so long as you use the right type of pen, that is. Rocketbook's creators encourage users to write in the notebook with a special pen that uses a special kind of thermo-sensitive ink that turns clear when exposed to high temperatures. Rocketbook is designed to be completely microwave safe, so unlike other notebooks, it won't melt or catch on fire when you nuke it.

Okay, that may be a step too far for most of you, so feel free to continue to use a normal book to retain all your information until you get comfortable with other concepts, but it does give you an idea of the impact technology could have on our lives in the near future.

When doing your research, learn how to search properly on Google. It astounds me that most people don't know how to even do that yet, so they miss out on a treasure trove of information. Not sure how to do it? Well just Google it and find out. Actually, now I'm starting to sound like my daughter, who asks why she should have to study when she can just Google the answer.

Once you have the general theme in mind for what you would like to do for your product, idea or venture, you then have to plan the content. One good way to help you do this is to go to www.amazon.com and find books that cover the same topic. You could look at, say, the top ten books in your niche individually, by opening the description of each and looking at the table of contents. These chapters are the various aspects that people are interested in within your topic, so you can develop your offerings around similar concepts. Actually, you can take that a step further and check out the comments in the review section for those books. Often people comment on what information they would have liked to see, which provides a great insight to help you develop your idea and content.

So, just starting mapping out the topics, concepts, and talking points for your chosen topic, whilst at the same time looking at which format suits best.

Digital Items

When goods are digital they can be delivered with perfect quality at nearly zero cost, and they can be delivered instantaneously. There are three kinds of digital products:

1. Written.
2. Audio.
3. Video (either screen capture or live on camera).

Where do your interests lie? No need for a decision just yet, we'll expand on this shortly.

Alternatively, you may not fancy writing or talking, and prefer to deal in the world of physical goods. So how can you up your game doing this? By using technology of course, but also by leveraging off the biggest and the best, particularly either Amazon or EBay. Another good option is www.shopify.com, which is one of many options for running an online store, using their all-in-one ecommerce solutions. Other options include Big Commerce and Volusion, so it's good to know you have choices that can make the transition to an online business a trouble-free experience. There are, of course, costs involved for utilising their services, but it is far cheaper than having a retail store!

Amazon doesn't just sell books, they sell nearly everything, to the tune of US$88 billion in sales through their online platforms last year. They continue to show great growth year after year and you can set up your own account to sell products at www.sellercentral.amazon.com to get a slice of that pie.

Now you can sell your own brand of products there, competing with the big names. Why would they buy yours? Well, you're leveraging off the credibility of Amazon; people trust things there. Alternatively you can source thousands of suppliers from around the world, who manufacture products that are currently selling well, but will do them on private label. Thus they become your brand. Other options include sourcing products from somewhere like www.alibaba.com (China's huge ecommerce business) where wholesale products are available at very competitive prices. If you haven't heard of alibaba.com yet, you should check them out. In just one day last year, November 11, 2014 they had online sales of over US$9.3 billion. Yes, in one day.

You can promote your sales on Amazon's keyword ranking and get near the top of the page to be seen by millions of people. There is a process to doing this successfully, but if you prefer physical products, this can be a great option.

Some people reading this may not want to sell physical products, nor is it their ambition to make money or change their lifestyle. They may simply see this as their opportunity to do some good. There's every chance they have been passionate about some cause, say for example, animal welfare or rescue, and see this passion as a platform from which they can now make a difference. They can now forge ahead and communicate with a tribe of like-minded people, they can passionately impact others in their community, or they can utilise technology to assist in making it easier to do the right thing by animals.

I should be clear that whether it's physical or digital products, I spend a lot of time with my clients in this stage, as devising your creative content or physical product business can be make or break. Deciding what the message is that you want to deliver is crucial—it has to resonate with people, but still be clear enough to differentiate yourself from others. Once you're ready, you can start to prepare your product.

Mark My Words

- Pleasure or pain—solve one of these.
- Good ideas have a blend of empathy and imagination.
- Change your perspective on things.
- The power of the Internet is the leverage or exposure it provides.
- Look for a well-defined niche (niches mean riches).
- Know your 'why' and it will make you cry.
- Be authentic and let your passion shine.
- Energy is the key to producing great output.
- Never undervalue your life experiences.
- Profit comes from externalising the passion.
- There's always a way to be innovative.
- Utilise digital goods—written, audio, video.
- Create a product so you're not trading time for money.

Chapter 6
Getting it Together

"Take that one idea. Make that one idea your life—think of it, dream of it, and live on that idea. Let the brain, muscles, nerves, every part of your body, be full of that idea, and just leave every other idea alone. This is the way to success."

—Swami Vivekananda
Indian Hindu monk and chief disciple of the 19th century saint Ramakrishna

Written material, as in a report or an eBook, is the easiest to create and a good starting point for many new businesses. In fact, a lot of Internet marketers use an eBook as their free giveaway to help them build a list or database of people interested in what they have to offer.

Let me digress for a second. I'm sure you've heard of the saying 'the money is in the list' where Internet marketers infer they can make money simply by selling to their existing database of contacts and clients. It is true of course, but here's the frustrating thing: it's not only for Internet marketers. I've seen a statistic that says 87% of business for a mature business comes from existing clients. Obviously that's just generalising, but think about your business. Would that percentage be about right? That means that the large majority of your client base will be repeat customers, if you treat them right. But—and it's a big but—do you know who they are? Do you have their name, email, and text details? If not, you're missing out on a huge opportunity; the chance to stay in contact, to inform them of specials or upcoming events, to make them feel valued, and to stay at the top of their mind so they're not tempted to go elsewhere.

Think about this for a minute. I'm sure you have a favourite restaurant or two, but do they have your contact details to keep you informed? Imagine the power of a message informing you about a new item on the menu to tease your tastebuds and stimulate you to return. Or, here's another great example: you might go to music concerts or football matches, along with many thousands of other people. Wouldn't this be a great opportunity for them to capture your contact details and send you relevant information? For example, if they know you love ACDC, then a quick email highlighting their branded T-shirts for sale could have a beneficial outcome for all parties. Similarly with your favourite football team, imagine an email with details of their new kit and you have the chance to buy it before it goes on general sale to the public. Bang. Hundreds of thousands of dollars in sales just from a simple email.

So there's no reason why you can't do something similar if you have an existing business. Treat your customers with respect, but get their details as part of your service to them. That's the best way to add value.

Now those of you who are in professional services will already have your clients' details and know the value of marketing to them. For you the message is more about sharing your knowledge to build your profile.

Being Published Gives You Credibility

You know what's amazing? I used to believe it was smart to keep your ideas and your information to yourself. Why give it out for free for everyone to pinch, because then your clients might think they don't need you? But the reality is that the more content you provide, the more influence and respect you earn. You are perceived as someone of authority and people want to do business with you.

Better still if you publish a book. Why? Because it positions you as an expert, a person of standing, a person of influence. Aside from these external matters, it can help from an internal perspective—it's also a good way of challenging yourself to keep your cognitive functions in peak performance.

In most businesses, but particularly if you're in one of the industries under the professional services category, then you have to establish some credibility to enable you to compete in the marketplace. If you're in the business of providing advice, then people need to believe that you are an expert in whatever you do. A great way to do this is through publishing a book (you'll notice I didn't say writing— it's the end result of publishing it that makes a difference).

Books work for you. They just make sense. How do I know that? Well, you're reading this, aren't you? They help you spread your message and give you that point of differentiation. They give you credibility.

With advances in technology and individual publishers these days, the costs, and the whole process of producing a physical book, are much less than just a few years ago. You will no doubt have heard of self-publishing, the ability to do it yourself instead of the more difficult way of proceeding through a traditional publishing company. Self-publishing, or as it now being referred to as independent assisted publishing, has become a very popular trend amongst first-time authors. This enables you to retain a higher percentage of the revenue generated, and gives you more say over the content, whilst you still retain the digital rights and have more control over the destiny of your work of art.

There are basically three ways to print the book— on an offset press, a digital press, or via print-on-demand. You'd use print-on-demand to save on the costs of shipping quantities of books all over the world. Otherwise most of us first-time authors will use digital printing for our initial shipment of books to be sold locally. While the unit price of each copy printed is higher than with offset printing, the average cost is lower for very small print runs, because setup costs are much higher for offset printing.

There are a number of different models you can use to print a book, so you need to determine which is best for your circumstances. As a self-publisher you can organise everything yourself such as typesetting, ISBN, barcodes, book cover design, and printing, which costs anywhere from $500 to a couple of thousand dollars. This gives authors the greatest opportunity for profit as it is a fixed cost to print and the author receives the entire cover price from any sales generated, minus distribution and marketing costs.

Then there are full-service companies that do everything for you, but you pay for it. This will cost you more, but may be worth exploring as at least this gets you into the market and you will find value in their expertise through the publishing process. Check out www.reedsy.com, who are UK-based, or in Australia try www.inhousepublishing.com.au.

There have certainly been enough changes in this area that I thought it warranted speaking with an expert to clarify the benefits. So I posed four questions to Ocean Reeve (yes, what a cool name), the publishing manager at InHouse Publishing, a Queensland-based company that assists authors through the whole process.

Mark Brigden: *Technology has certainly opened up the potential for anyone to write and publish their own book. Can you explain how digital and print-on-demand technology have changed the traditional printing and publishing model?*

Ocean Reeve: *The biggest factor in this area is that now both independent authors and large publishing houses can produce a run of books at a much more economical cost. In the past, at least 2000 books would need to be produced to make a book run feasible, and in the case of traditional publishers this was often a high risk, especially with an unknown author. With the growth of online purchasing and publishing it meant that risk was even greater and many publishing houses over the millennium transition suffered greatly with the change that was occurring. Nowadays this is more of a calculated risk and a run of 500 or 1000 can be produced to test the market and ascertain demand for the larger runs.*

Where it has really made an impact is with independent publishing. The digital production developments and even print-on-demand models have enabled the authors who more than likely wouldn't get a look in with the big boys, to now publish and print/distribute as few as one copy to 1000 copies economically, and successfully market both nationally and internationally.

Print-on-demand does have its negatives in regards to price, and also in quality. Whilst acceptable, there are more variables in the quality control with POD than in digital or offset production. However it is a developing area that will streamline even more successfully in the future.

Mark: *When would people go to one of the large publishing houses these days?*

Ocean: *My advice is, don't. Build popularity and success with independent publishing and draw attention to yourself for traditional publishers to contact you. This way there is more negotiating leverage when locking down contracts. However, it is through literary agents that you have the best chance of success.*

Nothing has changed in the pitch to traditional publishers. Whether you have a synopsis and a few chapters or a completed manuscript, they still expect it to be pitched in the same method, and you as the author will still need to wait patiently for a response. At what stage should this happen? Following a professional appraisal, peer group feedback, and self-edit would be the time to submit a manuscript for consideration.

Mark: *How does assisted independent publishing work?*

Ocean: *You have traditional publishing and self-publishing. Assisted independent publishing is in the middle, where the traditional process is implemented within the self-publishing process so the author still has* *to invest in the project but has final editorial say and input into design in line with the expertise a professional publishing consultant brings to the table.*

While it is primarily the author's responsibility to market and promote the published work, assisted independent publishing requires a commitment from the publishing house to also guide the author in this area, build awareness, assist the author, and ultimately sell copies of the author's work where possible. Some, but very few, also assist in order fulfilment and distribution.

Mark: *I imagine ownership of the digital rights is very important, but what else needs to be taken into account?*

Ocean: *Trade standard is ultimately the key factor here. Ensuring that your book meets the industry standards to be widely accepted into the marketplace is essential. This means ensuring that a professional appraisal is undertaken and followed up on, that a thorough copyedit and structural edit is completed, and the book is proofread at the correct stages of publishing. It means that the format of the book fits with what is expected by the retailers and readers, and that the cover design is effective from a promotional viewpoint. Prior to all this, certain legalities need to be considered, such as liability, possible defamation, copyright, fair use, etc.*

While the process of bringing a manuscript to trade standard can be quite daunting, it is

> *manageable, and essential that it is addressed, as it can affect not only the quality of the work, but also potential headaches further along in the publishing and marketing process.*

Whatever process you decide to utilise is up to you, but if you have the time to dedicate to making it happen, nothing spells authority quite like publishing your own book.

Remember, in most cases it's not the book that will be your ticket to riches, it is the other doors that it opens, because you are now perceived differently.

Once you have a hard copy organised you should also turn it into an eBook, which are growing in popularity. They do have some great advantages, such as:

- Storage—people can store thousands of books on one device (they're like a mobile library).
- Easy to buy—purchasing an eBook is just a click away.
- Accessibility—when you finish one book, you can easily obtain the next at many online outlets.
- Price—the price of an eBook is usually less than a print book.

To construct your eBook, many use an aggregator such as www.smashwords.com, a free service and the world's largest distributor of independent eBooks.

They convert books into the most common eBook formats for eReaders such as iBook, Nook, and Kindle. Smashwords charges a commission of around ten per cent, while competitor Apple charges thirty per cent, so check it out for yourself.

To some people the thought of sitting at a computer and writing a book is incredibly daunting. However they love talking about their concept, so they can structure their offering around their talking.

The Spoken Word

Audio can be a CD, an audio recording, or a podcast that you create and others can download. Some people aren't great readers, so prefer to receive the message through the spoken word. It also gives them the flexibility to listen to it whilst they're doing something else, e.g. driving to work, sitting on a train, or exercising. Plus, to many people it gives their imagination more freedom to roam, and helps stimulate their creativity. In some respects audio products are perceived to have more value than the written word, so you can charge more for these products.

Some of you may not be familiar with a podcast. It is a digital audio file made available on the Internet for downloading to a computer, iPod, or iPhone, usually available as a series, new instalments of which can be received by subscribers automatically. A podcast allows you to connect to your audience on a personal one-to-one level at a time that suits them. They download it and play it when they want. One of the most common podcast subscribing tools is iTunes, where there are thousands of podcasts available on a wide variety of topics. There are some fantastic ones available with quality information, so you should certainly be listening to some as part of your own self-development.

Years ago I hosted a podcast with a friend of mine. It was based around football—well, mainly the English Premier League—and was just a creative outlet as we didn't think to create products to monetise at that time. However the take-up rate was incredible, with thousands joining every week. There was no marketing, just word of mouth, yet it quickly grew to become the 7th most popular sporting podcast worldwide, ahead of even the BBC London.

In terms of exposure, an Internet podcaster can potentially reach a much larger audience than any traditional radio personality ever could. With potential numbers in the millions, it's easy to see why there is so much hype behind the technology and many business folk eager to get behind it. How could you take advantage of this? How could you use it to be regarded as a person of influence within your community, or tribe of like-minded people?

Let's look at an example. Say you're a real estate agent and you're looking for a way to stand out from the crowd. You can, of course, create a podcast about real estate, but a Game Changer would create a podcast about your local town, or state. That way you could interview all the movers and shakers in various industries, which would enhance your reputation as a leading businessman in the community, and it would be of interest to a much wider demographic. Of course you would always find a way to slip in something about real estate. Or you could look to interest people nationally by introducing some other aspects to give it a more holistic approach. Imagine if you then had a product you could sell, such as an eBook, after you have built your credibility with your audience. Too easy.

Just on that point, you can now contact people via Skype and record the interview with Call Burner. You could also record interviews via www.freeconferencecall.com and download it as an MP3 file, or even buy an app called Tape a Call Pro which allows you to record conversations on your iPhone. Then, an easy way to transcribe it is to go to www.idictate.com and pay an expert to do that for you. You then have content for your podcast, which can also be used as the basis of your eBook, for a nominal expense.

Get it on Video

Video can be delivered in several formats, e.g. through a DVD, a clip on YouTube, or even via a webinar.

Instructional DVDs have been around for many years and may no longer be the most cost-effective medium to utilise. However, having said that, if you are an expert speaking at a seminar, then it makes sense to package something up for those who aren't able to be there, i.e. create a product that you can sell afterwards. Production costs need to be factored in, but there still remains a market. Attendance at the seminar may have been free, but you can still charge a premium for everyone else to be able to access the information.

Online video is growing exponentially, with over four billion videos viewed daily. If you use YouTube for your business you can easily reach your audience, both by creating your own videos and advertising on other people's videos. Video-streaming platforms like YouTube have become so big you're guaranteed to find a group of people who will become your raving fans and customers, as long as you educate, entertain, and provide solutions to their problems.

Not sure where to start? Don't worry, YouTube has some comprehensive instructions via their Creator Hub and Creator Academy sites. You can use these videos to professionally promote your product or services, plus, once you generate a following you can also earn cash from advertisements on your site.

Alternatively you can utilise one of the online learning marketplaces, like www.udemy.com or www.lynda.com (now owned by Linked In) or for technology products there's www.trainsimple.com. The Internet isn't just a powerful tool for communication, it's the most potent force for learning and innovation since the printing press. ELearning is certainly booming and the combination of online and video just increases the demand. I mentioned the success of one of the courses in the introduction of this book, but how about this for a statistic—the ten most popular teachers on udemy.com have collectively earned in excess of US$8 million, and it only started in 2010. That's serious money. If we look at another site, lynda.com, they have provided video training courses to over four million people already, although they invite their instructors to participate, so you need a high profile to get a gig there. (Maybe write your book first.)

There's no doubt that people like the flexibility, ease, and comfort of paying to learn something online, and this is a huge growth market. Online is a less expensive and more convenient way to deliver this type of content.

I'm not suggesting that you will reach the same heights in selling your training course, as they've had a head start on you and first mover advantage helps them, but it does show the potential. What special niches could you target? Remember, the more focus you've drilled down into a smaller niche or subsection, the easier it will be to own that space. So you want something that's not mainstream.

When you are deciding on a topic for one of these training courses, the same criteria of passion and pain apply. People will pay to learn about something they are passionately interested in, similarly if you address a problem and show them how to solve it. These courses require a high video content, so you have to be sure you can demonstrate something on screen, or some other action, as just you talking for a few hours may not have the desired impact. Again, you have to spend some time initially on course development, but once it's available you just have to drive traffic to it.

An example of a good course might be 'Mind-blowing Results with Social Media for Finance Brokers'. You have targeted a specific market and have enough variety of potential topics to make it effective viewing. Hmm, might try that one myself once I've finished writing this book!

Traditionally, technology courses have sold well, but that may just be because the early adopters of this style of learning were technology geeks anyway. More conventional topics may soon be in demand as the general populace warms to this style of learning.

Udemy.com have different income streams. For anyone that you introduce (say via your website or an advert) that buys your course, udemy.com take only a 3% transaction fee for collecting and processing the payment, and you get the balance. Alternatively if they introduce the buyer, then they retain 50% of the prescribed course fee, but the remaining 50% is better than nothing, considering they generated the possibility. That seems great to me; let them do their marketing and take half the proceeds. Sure, there is a bit of work to do in terms of product development, creation, publishing, and marketing, but it can be a lucrative way to gain exposure, revenue, and industry credibility. You can of course distribute your course for free, particularly if it's a medium to build your credibility and/or sell other products.

By the way, many of the eLearning platforms use technology developed by Moodle, an Australian company. Moodle's user statistics are astounding, with their website reporting that more than 45,000 learning communities take advantage of the software in over 205 countries that speak eighty languages. That all translates to a user base of thirty-two million people.

There are of course any number of sites you can utilise. Alternatively be on the lookout for new movers. If someone like Amazon moves into this space, drop everything you're doing and create your courses, as they will take eLearning mainstream and you want to be on-board for the ride from the start. That tip alone could be your retirement fund. (Yes, I'll accept cheques or paid holidays to Hawaii.)

Web Seminars or Webcasts

Another spin on video is used by an Australian guru, Steven Essa. He uses webinars to promote his products, but he really is a Game Changer. He will start any process by promoting a webinar, so he'll advertise via social media or joint ventures with other people who have a list, to attract people to his webinar. He suggests you could all do this before you have even built a website or designed a product. In the webinar you tell the people what you are going to provide, sell the concept, and take the money upfront before you even design your training program on whatever your niche is.

In addition to that, the interaction with attendees when they ask questions provides you with a list of issues to address in future webinars.

Webinars produce higher conversion rates, reported to be as much as ten times higher than website sale rates. One of the big reasons for this is that the power in a relationship changes. Cold calling or static information gives the customer the power, however if they opt in to a webinar where you are holding court, you're held in higher regard. Webinars also give you the chance to use adding-value techniques and interact with attendees to overcome any doubts or questions they might have.

Here's a tip from the master himself, Steven Essa:

> To generate sales thanks to webinars, it all comes down to the idea initially. If the topic is interesting, the information is not already widely available for free, and the target market is hungry for that information, you know you're onto something. In addition, the presenter has to be articulate, confident, and convey passion in his voice. When it comes to presenting the offer at the end of the webinar, the overall package has to be priced correctly—not too cheap and not too expensive, and the product needs to make sense. The product always needs to feel as if it's a continuation of the webinar and not just a random product.

Essa has fully embraced the online mantra, confirming:

> Thanks to technology, all our staff work remotely from home, everything in the business, from selling, to delivering the product, hiring staff is all done online. We promote our webinars on social media and via email marketing, then we deliver our products and services using membership sites designed for our customers.

If you get the chance to see Steven in a seminar you should certainly take the opportunity, as he will open your eyes to the potential and the benefits of introducing webinars in your business. He will demonstrate the impact to you by selecting someone from the audience, and within a couple of hours have them host their own webinar, making thousands of dollars on the spot. And you'll love the fact that he stays authentic to his true self and doesn't try to be a slick salesman; he just tells it like it is.

I strongly believe that webinars will be the future of marketing. They allow you to present an in-depth presentation using live video, slides, and screenshots, whilst building a closer relationship with attendees. Through the chat function you get to answer their questions on the spot. It provides a personal aspect to marketing and brand awareness.

What could you use a webinar to deliver? The options are wide open, as anything of interest can be distributed via this method. In addition to this you could set up a business model where you partner with a specialist to distribute their products. Let me give you a simple example. Think of any expert, no matter their occupation, that works for themselves or has a small team. They no doubt have a busy schedule, but it's trading their time for dollars. They don't have time to spend on creating products that could generate passive sales for them, nor to do their own online marketing. This is where I, or you, could partner with these experts. They have knowledge that a much wider audience than their local clients will be willing to pay for. So we set up a series of webinars for them, automate it, distribute an email to a database, hire someone to build a membership site, and arrange some social media coverage, all in partnership with them. They don't have to spend time preparing this and just have to focus on preparing their presentation, with your help. Their product or service, which could be just a series of webinars, then gets sold online. You have a partnership, splitting all profits 50/50, so it's a win-win situation.

I use WebinarJam Studio as my delivery platform and will utilise that technology to deliver some crazy sessions in our membership site. The flexibility and professionalism it provides basically allows you to run your own television channel; your own personal desktop broadcasting service. It also allows you to store your content for replay at your customer's convenience. And if you add their Karta Professional shopping cart, with CRM and the ability to publish, track, and survey along the way, you have the full suite of Internet marketing products at your disposal.

So there's lots of different platforms or styles for you to consider. Again, it comes down to your preferences, plus what approach you believe will have the most impact on your target market. You don't have to choose one, you can adopt a multi-pronged approach, but it makes sense to master each one before moving on to the next.

Remember, to reach potential new clients you have to have some influence outside of your direct sphere. The way to do that is through utilising technology, be it video, a podcast, or whatever form of social media you choose. Use it to express your opinion, demonstrate your knowledge, explain your expertise, and communicate your beliefs. That way others will find you.

Strategy

It's great to have a big idea and dream of turning that concept into dollars, but of course you wouldn't start any business without a strategy and it's exactly the same online. You need to identify your target market, work out how to reach them, and what products they might want.

Again, don't make the mistake of thinking there's already someone doing whatever your idea is. Look at television as an example. There were a number of free-to-air TV stations when Foxtel introduced a subscription option. Now with TV advertising dropping (thanks to online advertising taking its place) and less people watching the adverts, the model was changed to one that builds regular recurring income. Stan and Netflix and other competitors have jumped in. So can you change the delivery mechanism within your industry to generate this recurring income?

Remember, when you want to do something different, you have to change the way you think. Look at it from another angle. Try a fresh approach.

The strategy you use to deliver your work will be a personal decision. There are a number of options as we've seen, so the decision comes down to whichever you think will give you access to your target market. There is no 'best' strategy, as things are very fluid with technology, and you have to align your strategy with your interests and your goals, both business and personal. This can save you much stress, floating between all the different options for your online marketing activities.

Remember, when you're clear about your goals, this clarity keeps you focused on the stepping stones along the way. As you'll see when you create your action plan, your strategy is an integral part of your roadmap.

One thing to remember is that your website should be more than just an online brochure. It should be more than just an outlet for branding your business. Business only happens when you make a sale.

I heard a great saying, 'most websites are like a billboard in the desert, not many drive past them'. So you have to first attract people to your website, and once you have their attention you have to make it count. It's important to integrate a sales process. People are there now, so help them buy your goods or services, or at least build a relationship with them to start the journey towards a purchase.

The way you keep track of all this online is through a sales funnel. As they move down the sales funnel they pass through categories such as leads, targets, prospects, qualified, and finally clients. If we look at this in a simplistic manner, you want to get traffic to your website, or lead page, capture their details, and then build a relationship until they buy your product. Depending on how good you are at targeting your ideal customer, you will no doubt get a large number of people just looking—tyre-kickers—who are unlikely to progress further down your funnel. Attract the right people, through adverts or links that they think are written just for them, and they will quickly pass through your funnel. Most sites use a series of structured emails to build this relationship and progress these contacts towards the purchase decision.

If you're not working alone, or you're using others to help you work on selected projects within the business, take advantage of technology to ensure everyone is on the same page. Lack of communication has caused many a problem, so make sure each step, the person responsible, and the time frame are recorded.

You can use a free application from www.trello.com to record lists, allocate tasks, and ensure management of each project is maintained. Another great option is www.asana.com, a free online management tool to help you keep your team organised, connected, and focused on results. Alternatively, you could also look at another web-based project management app called Basecamp.

Some of these things might be new to you, or you might have been an early adopter and are now proficient with these technologies. At the top level of town though, there are some big changes happening …

Mark My Words

- Treat your customers with respect and get their contact details.
- Keep them informed.
- Being published gives you credibility.
- Independent publishing opens up opportunities to anyone.
- Podcasts are increasing in popularity, reach, and influence.
- Use video to build a relationship with people.
- Webinars are the future of marketing.
- Align your strategy with your interests and goals.
- Business only happens when you make a sale.
- Take advantage of technology to coordinate your team.

Chapter 7
The Big Picture at the Top End

"In order to properly understand the big picture, everyone should fear becoming mentally clouded and obsessed with one small section of truth."

—Xunzi
Chinese Confucian philosopher born 310 BC

Don't yet believe the world is different? Consider this. In 2015:

- Uber, the world's largest taxi company, owns no vehicles.
- Facebook, the world's most popular media owner, creates no content.
- Alibaba, the world's most valuable retailer, has no inventory.

- Airbnb, the world's largest accommodation provider, owns no real estate.

And that's now, with many changes still coming. In the US, ten per cent of all business is now conducted over the Internet, with the percentage likely to be similar across developed countries, and continuing the increasing trend.

So what's shaping the future? The Internet continues to develop rapidly and has broadened our horizon of opportunity. Moore's Law still holds, as processing power and capabilities continue to double every eighteen months. The increasing capacity of devices will continue along with an exponential increase in the speed of data transfer. There's also increased connectivity between devices with access moving from personal computers to mobile phones.

The industries impacted the most to date are those with informational content; the service sector industry in general, but more specifically, media, culture, and entertainment. As Simon Cowell said:

> For the music business, social networking is brilliant. Just when you think it's doom and gloom and you have to spend millions of pounds on marketing and this and that, you have this amazing thing now called fan power. The whole world is linked through a laptop. It's amazing. And it's free. I love it. It's absolutely brilliant.

Other industries will be similarly impacted as the technology develops. The big gains will be in biotech and life sciences, but technology is also going to change many everyday things. I'm sure you've all heard about the driverless cars that are being trialled at the moment. It seems far-fetched of course that wide spread usage will eventuate by 2025 as some experts claim, however, look at the potential benefits. There are currently 1.2 million road deaths every year across the globe, the large majority of which are caused by driver error. If you can make a serious dint in that number, that would be a huge benefit.

Let's look at other things. The hottest investment process will soon be something called Social Sentiment Investing. That's where key words/products are tracked in social media, where word can spread like wildfire amongst people with similar interests or problems. Thus you can identify trends for demand amongst different products, and it makes sense to use this sentiment to anticipate sales (and thus possible share price movement) of particular products. Think of it as a database of human intentions. This technology is called neuralytics and will form an integral part of the marketing process for big companies. It enables them to use big data, machine learning, and artificial intelligence to identify sales patterns. Another way to describe it is predictive analytics. So linguistics and data science will be the next generation of marketing tools, used to identify purchase intent, and lead detection and sentiment analysis.

The Big Data movement will change the way large corporations function, with the intent being a better understanding of the consumer. So it's not just the aggregation of the data, it's the interpretation and insight that can be derived from it. This requires large databases of personalised information with specialised software performing analytical tasks to identify trends, patterns, and groupings. Ninety per cent of the world's stock of data has been generated in just the last three years. This creates a huge demand for data scientists or those that can work with numbers to derive meaningful assumptions that will drive future corporate actions.

Merchandising will change too, with one example being eInk integrated into garments, shoes, etc. Think of an eReader, which uses the same technology, and the impact they have had on the book market. Using this same technology will allow greater personalisation of the clothes you wear and the messages you want to send to the world.

Or if that's too far-fetched for you to grasp just yet, how about a business that sells DIY kits to print custom designs on T-shirts, bags, and other surfaces? Check out www.lumi.com where a then-twenty-four-year-old woman started her business three years ago, after finding a method for printing on natural surfaces with a negative film, developed by an inventor in the 1950s. Jesse Genet spent two years trying to track down the inventor, and ended up convincing him to sell her the trade secret formula for $50,000. Kickstarter helped her raise $270,000 to get the business started, and now her product is sold in retailers as well as online. Her profits this year are over US$2.5 million. Now that's a Game Changer.

The way we raise money to get many businesses or projects off the ground has also changed. Kickstarter is a crowdfunding platform, which in the first six years since launching managed to persuade 8.88 million people to pledge more than US$1.79 billion funding 87,483 projects (as at mid June 2015). Crowdfunding is a way of funding a project or venture by raising monetary contributions from a large number of people over the Internet. There are two methods for enticing people to fund your project, either rewards (e.g. a copy of the book) or equity (allocating shares in the company). Certainly if your project has an artsy or humanitarian feel to it, this may be one way for you to investigate funding the costs.

Then there's 3D (and even 4D) printing, which will have a huge impact on the manufacturing industry, in time. Production is being brought closer to the end user and thus distribution and supply channels will suffer, along with the manufacturers. Production will be localised and customised. There is an opportunity for professional services around 3D printing, ranging from new forms of product designers, printer operators, material suppliers, all the way to intellectual property legal disputes.

Those of you who already have a business that sells products online through eBay or similar, should already be using Terapeak for the research data and sales tools that it provides. These days there's no excuse for not being fully informed. Terapeak crunch the data that enables you to have a better understanding of your marketplace. Remember, a Game Changer uses technology to gain an advantage, so use these tools before your competitors do.

You'll no doubt hear a lot in the near future about the Internet of Things, or machine-to-machine communication—the network of physical objects or 'things' embedded with electronics, software, sensors, and connectivity to enable it to achieve greater value and service by exchanging data with the manufacturer, operator, and/or other connected device. This means everyday objects will be capable of receiving, generating, and sending information.

With so many electronic devices in our lives linked together for communication, we have networks that will allow the industry to completely automate complex processes at whole new levels. Harry S. Dent, the economic forecaster, pointed out some potential benefits:

> Currently, biotech researchers in traditional labs sit at benches handwriting detailed parameters such as temperature, pH, oxygen, and content, tediously moving and labelling hundreds of small volume samples at a time. When these same processes are performed with robots and automation software, errors in execution virtually go to zero, and reproducibility goes through the roof.

> Labs across the country are starting to embrace the "Internet of Things" by employing massive facilities armed with robots connected to a complex infrastructure of lab testing software and databases. Everything is logged via barcodes and loaded in databases to even track inventory, allowing these facilities to replace low in-stock items right when they need them.
>
> Some labs such as Ginkgo Bioworks and the Emerald Cloud Lab are actually set up for outsourcing, meaning they allow external researchers to come in and use their facilities and expertly engineered (and cheaper) bio-organisms. That saves the researchers money by avoiding costly in-house research.
>
> This business model is very similar to how technology product companies leverage external chip manufacturers, instead of producing their own chips for their devices.

There are some big predictions about the ramifications the Internet of Things could have, however I'm sure there's still some major hurdles to overcome (e.g. preventing hacking) before it becomes mainstream.

The sports world will also undergo transformation. An example of this is American football where both college and professional teams have started using virtual reality to train their quarterbacks. They are fitted with a VR headset, allowing them to see a live-action 3D video replay of a football play from the quarterback's perspective, and can review that play from a first-person view over and over, looking in any direction. It's like real-time, real-situation training to help them prepare.

Aside from the large data-driven companies there are lots of changes at the smaller end too. Fiverr has been around for many years and is a great way to outsource a task to a qualified person in an overseas country for a fraction of the cost you'd pay in most developed countries. This covers a full range in graphic design, online marketing, video, copywriting, and most online services, so if you haven't used it yet, go to www.fiverr.com and check it out.

Uber has proven to be a disruptive force for the taxi and hire car industries. Can this business model be used in other industries, for example to cheapen distribution costs for the likes of Amazon? Could it somehow work for you if you had a retail business? The sharing economy will spread to other industries, so what spin can you put on it?

Globalisation has already impacted on many businesses, as products are now freely shipped worldwide, so it becomes much more competitive. The same thing is now happening with services. Need some help with routine tasks to free up your time? You can get a Virtual Assistant located overseas, but with the same ability to complete your tasks.

Heavy stuff, but it's not meant to frighten you. In fact a Game Changer would not be deterred by the upcoming changes; they would just realise that monumental change is in the process of happening, then they'd get on board, find a niche they can service, and ride the technology wave. When you can identify it's a time of change, that's when you embrace it and make money out of it.

Mark My Words

- Profound and monumental changes have already started.
- Most industries will be impacted.
- Business models are already changing as the sharing economy spreads.
- Globalisation has meant outsourcing to developing countries.
- Get ready for the Internet of Things.
- Now's the perfect opportunity to take advantage of this disruption.
- Crowdfunding could provide a solution.
- Look at social sentiment investing for your personal investments.
- Find your niche and ride the technology wave.

Chapter 8
Your New Online Vehicle

"Online, there's no time. It's always Christmas."

—Lewis Black
American comedian

Some people might like to make an easy entry into having a product available for sale and an online presence. So what's an easy way? Check out www.flippa.com, an auction site for selling existing online businesses. You may find one there that suits your needs perfectly, and everything has already been done for you. It's an excellent way to source a good business, and they display lots of analytics about sales to date, number of website visitors, and other relevant information. Alternatively, you may find a domain name that you think is perfect, as some of these are auctioned off as well. If nothing else, this can give you insights and ideas that you can use in your own site.

Most people will start off with a website and a blog that allows you to write about areas of interest within your niche. Check out the WordPress sites available that can offer a magazine style site, at www.wordpress.org (don't go to wordpress.com). This site make it easy for you to build one, and then you can add content to it as often as you like.

Should you choose to, you can even use free templates from www.weebly.com or www.wix.com. The artistic ones amongst you should even try out www.squarespace.com. So there's plenty of options, but still buy a domain name so that it sounds more professional. You need to be taken seriously. This applies not only to your domain name, but also to the look of your website. You will need some quality photos, and if you don't have any good enough then go to www.istockphoto.com or try www.shutterstock.com where you'll have to pay a nominal fee for a stock image. Or, try a freebie from www.freeimages.com or www.pixabay.com, a good site with high quality images you can use for any purpose thanks to their creative commons licence.

There are also a number of sites where you can buy impressive, short videos for your website, including www.videoblocks.com. This is a stock video marketplace with thousands of diverse videos selected from contributors. Actually that's also a great place to sell a video if you have already made your own. At the time of writing they will include any good quality video in their stock at no charge, and you receive full payment from anyone downloading it for their own use.

Many people make a living by just blogging, which is writing articles and providing insights into tightly defined niches. Yes, I've heard the argument that there is already a lot of information available on the web, but that's old school thinking. A Game Changer would realise you can amalgamate topical information into the one location so that those interested in this topic don't have to go searching. People are inherently lazy and time poor, so you do the work for them by aggregating appropriate content at your site or blog or in your product.

The more focused your topic, the increased likelihood of attracting advertising as well as paid subscriptions. Great, a dual income stream.

As more companies utilise websites and social media as a means of staying in contact with their clients, there's a growing demand for content writers. So building your own site first provides an excellent showpiece of your talents, and you may soon be in hot demand.

Of course, whichever business model you choose, it should meet the lifestyle business demands you decide are appropriate for you. Selling physical goods has different demands to selling digital goods. You can sell adverts on your site as an income generator (see www.google.com/adsense), or keep it clean and more focused.

Whilst there are several different strategies, most involve creating a product that can be distributed online. So who do you sell it to? Well, that's where list-building comes in. The ideal scenario is that you focus on your particular niche and target them through some cheap advertising with an offer of something for free, which they subscribe to. This can be an eBook, webinar, or audio product.

You will have heard the phrase 'Internet marketing' before. If you're not comfortable with it, it's just marketing done on the Internet to sell a product. There are three aspects that form part of the process:

1. How many people visit your website, either directly or through clicking on a link advertised elsewhere? This is called traffic.
2. How many of these bought something? These are your conversions, and as a percentage of the visitors it represents your conversion rate.
3. What is the average spend of these visitors? This is your earnings per click.

Mark Anastassi, author of *The Laptop Millionaire*, put this in simple terms for us to understand.

> Imagine a bakery on a busy street. If 100 people walk into that bakery, that's 'traffic'. If 10 of these 100 potential customers buy something, this means they were converted from a prospect to a customer—that's 'conversion'. Ten people buying, out of 100 visitors, represents a 10 percent conversion rate. If these 10 customers spent $15 each on average, the baker makes $150 in sales from these 100 visitors. That's $1.50 earnings per visitor.

It's the same process online.

So Internet marketing just looks to maximise each of these three things. When people give you their email in exchange for a free product or to subscribe to your newsletters, this is called permission-based marketing. You can then email them further information about your products and services. You know they have an interest in your topic, so you want to build a long-term relationship with them. Never think in terms of the transaction, always think in terms of building a sustainable relationship. And don't spam-email people who haven't asked for access to your information—it's illegal.

We touched on the term 'sales funnel' briefly earlier—it's just a metaphor of a funnel (wide at the top, narrow at the bottom) to monitor the sales process. At the top of the funnel you have 'unqualified prospects'—the many people who might need your product or service. They might work their way down through various steps, such as reading the content on your website. Then at the bottom of the funnel, several sales and delivery steps later, you have people who have taken delivery of all your products or services and have paid for it. There are a number of reasons why people may drop off throughout the process, so steps are prepared in such a way as to retain their interest as long as possible through to the purchase decision stage.

Lead Pages

Rather than just relying on prospects stumbling across your website by chance, Internet marketers use what are called lead pages or squeeze pages. People will respond to one of your advertisements and be directed to this squeeze page. This enables you to collect their name and email address for them to proceed. They then go to a sales page, which is just a webpage designed to sell your product or concept. It could include a video, some fancy marketing, and some testimonials. The next step is usually an automated email follow-up to build rapport with that interested party, and then a call to action so they don't hesitate and put off making the purchase decision.

Now remember that earlier we suggested the easiest way to do these things was to copy someone else. So how can you develop your sales funnel? Just find a product that has the process you like by going to www.clickbank.com and use the marketplace tab to find a good quality, but similar, product to your own. After you buy their product you'll have access to their full sales process (their funnel), which you can copy and personalise to your style. No need to re-invent the wheel.

To sell digital products online you need a way for people to pay you. This is called a merchant account and allows you to accept credit card payments (and also chargebacks and refunds). The first one to consider is PayPal. You can in time look at other options, such as

Stripe or Apple Pay, but most people are familiar with PayPal. Stripe is quickly becoming a strong contender and allows you to accept payments securely over the Internet with a similar fee scale to PayPal.

Apple Pay is an interesting concept that will probably revolutionise the way you purchase items yourself. It is one of many competitors in the mobile payments space, but has the advantage of utilising fingerprint technology via your iPhone to ensure security of the transaction. So the security is much tighter as no-one needs to see your MasterCard for every transaction—and security will continue to be an issue for all aspects online.

Check Competitors and Understand Your Options

We mentioned earlier that people respond to your advertisements, and that is what drives your traffic. But where do you advertise? Again, let's look at what has worked for others and not spend hours stressing over possibilities, or even worse, just reacting to opportunities given to us without operating in a strategic manner.

So who is doing something already successful in this regard? How do you find out what your competitors are doing that makes their advertising successful? Go to www.similarweb.com or www.alexa.com and put in your competitor's details. Near the bottom of the page, as long as your competitor is big enough, you'll see details of the sites that people have clicked on to get to their webpage. This means these are the sites they advertise on that drives traffic to their website. So you can now plan your advertising campaign around these same sites, which have been proven to produce results.

Before you get hung up on what may seem difficult to you at this stage, let's look at some possibilities within these online ventures.

Publicity about a court case in early 2015 involving two social media fitness gurus on the Gold Coast highlighted the type of income potential possible. Emily Skye promotes her photos and products mainly through Facebook, Instagram, and Twitter. Admittedly, she looks sensational in gym gear and posts inspirational messages about how she achieves her figure, along with links to exercise and weight loss programs. This has managed to generate sales of $15,000 a day. Yes, a day. The other fitness queen, Ashy Bines, had huge success with an online bikini challenge and also has a very large online following. No doubt they have worked extremely hard not only to get their stunning physiques, but also to develop informational and instructional products that they disseminate through the power of social media. Game Changers.

You don't have to take them on at their game, just find your own niche. It could be something simple that means something to you. For example, instead of trying to change the world with your activities, you could create a site that helps others who want to do this. You could provide them with ideas, templates, an avenue for promotion, and a community to join like-minded people in their pursuits. Or you could decide to pass on some old-fashioned values and have material available relating to good manners, etiquette, and community spirit. Whatever your area of belief or expertise, you first build a following amongst like-minded people, and then you find a way to build products around that concept.

You've also no doubt seen videos on YouTube and wondered how people make money out of that; well here's a great example. Michelle Phan started posting how-to videos on YouTube back in 2007 while she was in college, after readers of her beauty blog asked her for tips on how to apply makeup. She kept posting videos as she felt it would help her land her dream job at Dreamworks or Pixar, or at least give her an advantage over other applicants who weren't online. But something happened along the way; her following grew and by 2015 she had 7.6 million subscribers to her YouTube channel. She used this following to expand into books, a music label, cosmetics, and ecommerce, resulting in annual sales of US$120 million just seven years after she started. That's the power of online, the power of video, and the power of a Game Changer.

If you're just looking to drive your existing business further, or open up opportunities for yourself within your company, I'd look initially to utilising video. It continues to demonstrate growth and effectiveness.

You can use it in a variety of ways, including for sales messages and demonstrations on how to use your products, plus it presents a great avenue to showcase testimonials. It's amazing the impact that reviews and testimonials have on anyone considering their options—and this appears to be only increasing with time. As a society we want the validation that someone else has enjoyed the experience. Just look at sites like www.tripadvisor.com.au, for example, and you will understand the benefits testimonials can bring to your business.

Words are great, and video can be even better. Thanks to technology, video is certainly cost-effective and enables personalisation that words alone can't match. An added benefit is that it certainly helps with your Google search engine ranking, and we all know the benefit of being high on the list of search results.

If you've been in business for a while and are looking to ramp it up a notch, then it's not good enough anymore to just have a website. A static site is just a fancy business card. Include a video and people will relate to you. Then take that a step further and ensure your website includes a sales funnel—you have to monetise it whilst you have their attention. A sales funnel is basically the process a client goes through when purchasing products, from engaging their interest, to purchasing and follow ups. It's divided into several steps, which differ depending on the particular sales model. If you're not sure how to do this check out www.clickfunnels.com, an excellent site that provides templates to allow you to build your own sales funnel pages.

Videos save you time if you address repetitive issues, as clients can simply watch them rather than individually occupying your time. If you're short on sales staff, then put a demonstration on video and you just have to direct potential clients there, so you're maximising your available time. They are also a great way to record key moments in your business, short snippets that can then be edited together for a great highlight reel on your website. Once you have a good selection of videos you may want to even start your own YouTube channel.

Not sure about where to start with getting your videos up to standard? Just use software called Camtasia for PC, or Screenflow for Mac, which enables you to customise and edit your videos. You can record on-screen activity, add imported media, and create interactive content to give it that professional feel. You can get Camtasia at www.techsmith.com. By the way, another great product available from there is Jing. This software takes a picture or video of your computer screen and uploads it to the web, computer, or clipboard. If uploaded to the web, the program automatically creates a URL to the image that can be shared with others. This makes it a great way to email your clients with a link to a video. So for example if you're sending a proposal, this is much more effective than just sending a PDF document.

Let's look at another social media platform, Instagram. There's a great story about a swimwear company, Triangl. Started by two Australians in 2011, it struggled for a few years until they realised that selling via retail stores wasn't the answer. In January 2013 they stopped all wholesaling, adopted a vertical business model, and promised themselves to never discount, never go on sale, and never sell offline. Social media has been pivotal to building their market awareness. The pair started gifting swimsuits to bloggers and, thanks to their influence, word quickly spread online.

Today Triangl has more than 1.4 million followers on Instagram, thanks to American celebrities including Kendall and Kylie Jenner and singers Miley Cyrus and Beyoncé, who have posted photos modelling the snazzy, synthetic rubber swimsuits. Their original goal when starting out was to sell one bikini a day, but thanks to Instagram they sell over a 1,000 a day now, just two years after adopting their online process. Revenue this year will reportedly be about US$25 million. Lesson—social media sells.

Here's another example of a passion, a niche, and the power of Instagram. A Scottish woman with a passion for illustrating developed a ninety-six-page book called *Secret Garden*, comprising of elaborate black-and-white ink drawings of flowers, leaves, trees, and birds. The first print run was for 16,000 books, yet since 2013 they have sold more than 1.4 million copies. Who would have believed there was a target market for adults who liked to colour in? Her big break came in South Korea, where she has sold

430,000 copies, which kicked off after a posting on Instagram by a Korean pop star. So whatever your passion, whatever niche you want to target, this is a great time to get it out there.

There's a lot of talk about developing apps these days, and that can certainly be lucrative. Mobile business apps let your customers stay connected, whether they are online or offline. With over 8.5 million smartphones in use in Australia alone, and the increasing reliance on them, it makes sense to target this market. You can use apps to increase customer awareness, build brand loyalty, and drive sales.

Make Sure it's Accessible on Mobile

Close to 15% of global ecommerce turnover is already done on a mobile, and this transformation has happened in such a short period of time. Twenty years ago less than three per cent of the world's population had a mobile phone and less than one per cent were on the Internet. Now, two thirds of the world's population have access to a mobile phone and one-third of them are able to connect via the Internet.

An app (application) is a piece of software designed to run on your desktop or on a mobile device. While it may be hard to compete with a big outfit, developers who seek out and dominate niches outside the standard app categories are doing very well. Alternatively, look to associate your service with the expected huge growth for the Internet of Things (another tip that can fund your retirement).

You can even make money whilst giving your app away for free. This can be done through either in-app advertising, sponsorship, upgrades, in-app purchases, or bundle deals. There's always ways to integrate it into your business to enhance the service your customers experience.

There are already over a million apps available to be downloaded, so it is more than likely someone else is trying to deliver on the same idea you have. There's no need to be put off by this though. An app's success is not based on the originality of the idea, but rather how well it meets the needs of its users. So once again you can always refine a similar concept.

Something to think about: Flurry, a Yahoo-owned mobile app analytics platform, tracked the retention rates for Facebook Messenger and similar messaging apps and found them to be 5.6 times greater than rates for the average app. Messaging apps also have a daily use that's 4.7 times higher than the average app. The average daily use of an app across all categories is 1.9 times, whereas messaging apps are used, on average, almost nine times per day. So maybe include a simple contact mechanism in your app. A great example of this is Voxer, which provides walkie-talkie communication for your team.

You should of course check out the apps that are available. There are some fantastic products already out there that can be of benefit to you in your business, particularly from a productive viewpoint. There are apps to enable you to easily track your expenses, like

Concur and Expensify. Then there's the high tech ones like, say, Crystal, a new app that analyses public data to tell you how to communicate with some people. It creates a profile from publicly available information and suggests ways to approach people based on their character traits. Big brother stuff.

So whether you want a plain website, one with the latest design features, or a cutting edge promotion on a particular social media platform, there are a lot of choices to enable you to select the best method to reach your prospective clients.

One obvious method to gain exposure and credibility is to post a comment on social media. Don't fall into the habit of discussing your life or your issues, until you are regarded as a celebrity. People want to read things that will improve their life in some manner, not you whingeing or mentioning what you had for breakfast. You're a professional, so it's even more important to act it online, as people cannot interpret tone or intention. Look for areas to contribute to a discussion in some positive manner, which will demonstrate your knowledge and add to your influence. On LinkedIn for example, utilise the groups to find a common thread with your ideal target market. Remember though that 'likes' and 'follows' are not dollars. Don't confuse popularity with sales.

You'll often hear marketers talk about the benefits of engaging your clients. One cool way to do this every now and then is to do a survey through www.surveymonkey.com. It's free and when structured well can provide you with valuable information. Another great site to help you develop surveys, competitions, questionnaires, or basically develop any online form is www.typeform.com. This way there's no excuse for not looking professional on a small budget.

To help guide any potential prospects for your business, you should have a discovery process. Ask questions and find out what they want, how they operate, and which of your products would be of benefit to them. If nothing else, they might appreciate you taking an interest in them. This helps build that connection with them as they want to feel that you understand their needs and you're not just selling to them. For them to relate to you, you sometimes have to reveal your human side.

If they're aware of you, believe in you, and believe you already provide value to them, you can build a customer for life. The biggest loss of clients and potential customers in most businesses is caused by a lack of communication. There's no excuse these days for you to allow that to happen.

Mark My Words

- You need to establish your presence on the web.
- Internet marketing is just the online version of doing business.
- Model what successful people have done.
- Build your list of interested clients.
- Utilise video.
- Use social media to build awareness of your brand and products.
- Likes and follows are not dollars, popularity is not sales.
- Build a sales funnel.
- Don't discount, just look to add value.
- Ensure a solid flow of communication, hopefully two-way.

Chapter 9
Speculate to Accumulate

"Creative without strategy is called 'art'. Creative with strategy is called 'advertising'."
—Jef I. Richards

I'm sure you've heard of the saying 'speculate to accumulate'. It just means you have to spend money to make money. And I'm not talking about going to the horse races and gambling your hard-earned cash (although I do have a favourite betting system if you want to know about it). Here I'm obviously talking about spending money in business to make a return.

If you have an offline business, say a retail outlet, you spend money on numerous things. This could range from splashing more on the fit-out costs to ensure you create the right impression, to your marketing programme, which no doubt includes signage and advertising to build brand awareness.

Why do we do this? Because it's part of the lead generation process. The fit-out creates the experience so you want to attract anyone passing by with your visual impact, and ensure those that come want to return. Similarly, your marketing needs to generate some enthusiasm amongst anyone seeing your promotional activities to get them to want to experience your products. It's the cost of doing business.

It's exactly the same concept with your online venture, however the advent of technology has made it cheaper to have an impact these days. It's far cheaper to set up your online business than it is for a traditional shop front. It's far cheaper to maintain your business through ongoing fees, such as the cost for hosting and updating of software, than it is to pay lease payments and outgoings on your rental premises. It's far cheaper to use social media to advertise and promote your business, than it is to place an advert in a traditional magazine, Yellow Pages, or a newspaper.

War for Attention

It really is a war for attention. You're competing with other online ventures to grab the attention of potential clients and then hopefully build a relationship so they become and remain customers.

It's stunning when you realise you can potentially reach millions of people, and influence them, in just days, AND you haven't had to spend a cent on advertising. Of course, getting that kind of exposure is very difficult, even with the recent explosion of social media. That means your posts, or videos, have to stand out from the crowd. They have to entertain, inform, and excite. They must have a message worthy of being shared.

Too often we see most people trying to share as much as possible, and flooding their friends with posts thinking it's increasing their profile. It's not. People will just get sick of the rubbish and glance over your future posts, thus diluting your potential relationship with them. So don't listen to these Internet marketers that tell you to post, or send an email, three times every day. Post when you have something valuable to say. Post when you think it will make a contribution to the other person's life or business. Post when the content positions you as a person of influence.

Once you get experienced at posting and have time on your hands, you should then think of posts like you would if you were writing content for an article or a blog. It should be informative AND ideally include a call to action wherever possible, i.e. it should prompt some response. This could be:

- Sign up for a newsletter
- Download the report
- Refer to a friend
- Click through to sales page
- Watch a video
- Buy the product

With social media you can of course post a range of content with the intent of capturing the interest of people. They share it, their friends share it, and suddenly you have exposure to a large number of people. Posts with video are increasingly popular, and they don't need to be slick sales promotions. So you should explore your options through these free mechanisms before implementing an advertising campaign.

If you've dabbled in this already but need to improve, or you're just starting out, there are a few steps you need to take:

1. Understand where your existing accounts are at—in other words, conduct a self-audit, so that you understand what platforms you're on, how to access them, and what your passwords are. Make sure you are consistent across all platforms; use the same user name, profile picture, banner, and description (this is a key part of branding that is often overlooked).
2. Identify your competition and check on what they're doing—how often they post, what their content usually is, how many fans and followers they have, what they use to engage people (competitions, videos, giveaways, etc.), and whether they display reviews and testimonials. When you're on Facebook there's a great in-built tool that allows you to keep track of your competition. Just go to the 'Insights' tab and then under 'Pages to Watch' you can add your competitors' names, allowing you to easily see an ongoing

snapshot of their stats so you know what they're doing and how.

3. Set yourself some goals—if you're going to use social media then you want to maximise the results. Do you want to increase click-throughs to your website, or increase email subscribers, or even get more engagement than your competitors? Then write down some quantifiers such as by how much and by when. Any goals for anything you do need to be SMART—specific, measureable, achievable, relevant, and with a time factor.
4. Devise a contact calendar—how often do you want to post, who is responsible for this, what type of content (quotes; company achievements; industry news; tips; tricks; and advice—break it up so there's a good mix)?
5. Track your results—what platform performs best for your business? Should you increase activity on that, or increase the content on the other platforms to try to match the performance? Again, testing is the best way to determine what's best for your business.

If you do this right, in a professional manner and with a good spread of interesting content, you should generate a good following and greatly increase your exposure. However, Facebook and the others are looking to restrict the amount of free exposure you get as they build their revenues through advertising.

When Facebook changed its algorithms, the reach of organic posts dropped off substantially.

Advertising Online

Now what will the impact be of these posts if you add advertising, or if you pay to boost your reach, particularly when you have the power to target your posts to a particular demographic. The first advantage of this reduced cost in the digital world is that you can try things to see what works. Every management guru and business coach will tell you that you can't manage what you can't measure. So we test different things, from the types of ads we develop, including graphic layout and content, to which social media platform we advertise on.

John Wanamaker delivered a great quote years ago that summed up how most business owners felt. "Half the money I spend on advertising is wasted—the problem is I don't know which half," he said. Technology these days, however, enables us to measure our marketing programmes, thus placing us more in control. We can easily ascertain the success of our various promotional campaigns, including conversion rates and subsequent purchases.

Reduced costs and more control; this sounds appealing so far.

Another advantage is the leverage the Internet provides in terms of increasing your exposure and targeting your adverts towards your particular niche.

They no longer have to remember your details and drive to your store at a later date; they get instantaneous access to your offerings. It's almost like they're transported in time and appear in your shop magically. This means you have access to them while they are most interested. Compare that to a specialty magazine, such as one on horse jumping. Someone interested in your apparel may see your advert, but it could be weeks before they have a chance to visit your physical store and thus momentum and desire have been lost.

Bill Gates saw this coming when he said, "The future of advertising is the Internet."

We spoke in the previous chapter about building a profile by commenting in the various social media platforms. One important tip to remember, and ensure you focus on, is that it doesn't matter how many likes or followers you have if you're not driving them to your website or online store, where you can either capture their contact details or sell them something. This is the mistake that many people make. They share or develop their own pretty memes, but don't generate the link to their site.

You can of course also use social media for targeted paid advertising, to target a very specific market. On LinkedIn, for example, ads can be targeted by job function or title, company size, or industry and seniority.

Facebook is the common one to start with, so to run an ad campaign there you simply go to the Facebook

Ads tool, which you can access via www.facebook.com/ads or by clicking 'Manage Ads' in the drop-down menu on your Facebook account. With Facebook, you have many different ways of approaching an ad campaign, with benefits including:

- **Interaction**: Your ad is placed right on the homepage, which allows users to interact with it like they do any other piece of social content.
- **Reach**: Expand your reach to new potential customers who can interact with your content by commenting, liking, and sharing.
- **Followers**: Most brands also report a notable increase in followers through these social advertising options, since brand visibility increases significantly. For small budgets, you're likely to get the most bang for your buck with boosting reach. With your campaign you can customise who sees your banner advertisement according to:
    - ◊ **Age**
    - ◊ **Gender**
    - ◊ **Location:** From as large as a country to as small as drilling down to a local area.
    - ◊ **Interests**: Facebook analyses their interests, activity, the pages they like, and related topics.

◊ **Behaviours**: This covers things like their purchase behaviours, as well as which device they use.

You can actually get some benchmark data from Facebook to demonstrate the industry averages for categories such as the average click-through rate, the average cost per click, and the average cost per 1,000 views. To date, telecommunications and publishing get the most click-throughs, whilst clothing and fashion have the cheapest cost per click, probably as expected—but things change quickly in the online world. And of course you can set a budget of whatever is comfortable for your advertisement spending—say $5 a day to start with, and then see what the results look like.

Quality Ads

So what are banner adverts? They are simply the small rectangular advertisements that appear on web pages and can vary in appearance and content. A banner ad has a graphical element, so is somewhat similar to a traditional advert you would see in a printed publication, such as a newspaper or magazine, but it has the added ability to bring a potential customer directly to your website.

As you've probably noticed while you've looked at different sites, or even just your Facebook page, the actual graphic content, which is the creative aspect, varies considerably across banner ads. The simplest banner ads feature a static GIF or JPEG image, which is linked to your website. Another option is the GIF-animated banner ad, which displays several different images in succession, sometimes to create the effect of animated motion. Then there are rich media banner ads—these use audio, video, or Java and Shockwave programming.

If you want to create some of these basic ads yourself, try using the free software from design platform site www.canva.com. In fact, nearly half of the US Fortune 500 companies have been using them to prepare their business documents. They enable anyone to more easily create graphics, including slideshows; invitations; handouts; posters; infographics; Facebook covers images; cards; collages; blog graphics; and more. Actually, when you go to their website you'll find some great reading on their Design School pages. You can get some great information here, like the Complete Social Media Image Size Guide (a handy tool to keep).

Then if you like this creative side and want to do more of it, check out www.markerbook.net where you can get lots of free resources that will make your site look professional.

While you may have some talent, if you don't have time to learn you're probably better off outsourcing the creation of your banner ads to a professional, certainly at first. There are numerous sites for this, but a popular one is www.20dollarbanners.com. It's smart business practice though to not put all your eggs in one basket, and this is where your testing comes in. Try different banners to see which ones work best, try different places to advertise, and try different hooks. When you find the right mix, then push it harder.

Once you're online and generating traffic to your website through advertising, you can then step up your game even further through developing your network.

Mark My Words

- The costs of doing business online are substantially less than offline.
- Activities and promotions are able to be tracked, so tests should be ongoing.
- Use social media to extend your reach.
- Establish a social media content programme.
- Decide which platforms suit your model and ideal client.
- Consider adverts on the various platforms.
- Test what works and amend where necessary.
- Ensure your graphics are professional.
- Link your graphics back to your website.
- You have to win the war for attention.

Chapter 10
The New Networking

"If I had to name the single characteristic shared by all the truly successful people I've met in my lifetime, I'd have to say it's the ability to create and nurture a network of contacts."

—Harvey Mackay

Networking is part of the marketing and sales process. It always has been and always will be. It's a great way to build your profile, push you outside your normal comfort zone, and add to your business leads. People like to deal with someone they know and trust, so it's a perfect solution to gaining new clients without spending a lot of money.

However, networking is a skill that few people use well, and is now one of our most popular training courses helping you to articulate your point of difference in a manner that will make people keen to learn more and to deal with you.

Now if you can master it in an environment where leverage is also applied you will reap dramatic benefits. The online rules for networking use the same principles of reciprocity and strategic targeting.

The basics still remain pertinent—you need to focus on building a relationship with someone, not trying to sell them something. So this means as a starting point you have to have your personal statement succinct, in that it tells a bit about what you do, but leaves them wanting to know more.

As an example, if I was asked at a social function what I do, I could say, "I provide strategies and techniques to help business owners become outrageously successful, by giving them opportunities that their competitors don't have." Now I'd have their attention and I'm sure they'd seek clarification.

Then you need to find out more about them and display an interest in what they do. Gather as much information as you can politely. When you have their contact details you don't harass them trying to sell them something, you first provide something of value to them, because you're trying to build a relationship.

Always think about the relationship, not just the transaction. Customers for life are what you want to garner. I'm very big on getting away from transactional thinking, and instead building a solid relationship focus. The stronger your relationships, the less you have to seek new clients; they find you! Think of yourself as a farmer (cultivating a relationship) rather than a hunter (zeroing in on a target).

Thanks to these recent advances in technology, networking is now so much easier for you to undertake outside of your normal sphere of influence (usually denoted by your location). Whether it's a podcast, webinar, or Internet marketing in general, you now get the opportunity to bypass the gatekeeper. This gives you tremendous power if you use it correctly; for the first time in history you can access the decision-maker directly.

How can you take advantage of this? Never underestimate the power and influence of storytelling. What's your story? What examples or instances can you elaborate on to help push your cause or interest others? Communication is a key aspect. Successful leaders, be they politicians, sports coaches, or business gurus, are great communicators. They are able to be emotive and sculpt their stories to target your feelings.

Where to Network Online?

Now most people look to social media to do this in the online world, although many aren't doing a good job of it. Let's look at some of the basic platforms.

Facebook

Facebook is currently the most powerful social media platform available, with over 1.4 billion people registered as users. This should be the first place you get right as it allows you to market to your ideal clients. If you set up your page the right way you will attract your ideal type of client, which gives you the opportunity to build a relationship with them. Closed groups also give you the opportunity to lead your own tribe.

Google Plus

Google Plus encourages you to build new relationships based on your passion, interests, profession, or career. It offers a space to find information that's relevant to you and to share it with people who have the same interests as you. This helps you widen and build a new network of contacts. Google Plus also allows you to bring your brand closer to your followers and prospective customers in a new, unique way, much differently than other social networks do. There are over 500 million people on it. Importantly, if you're part of Google Plus, you're a part of Google, as everything you post on Google Plus is immediately indexed by Google. This positively influences search engine results.

LinkedIn

LinkedIn has 350 million registered users worldwide, and to most people this is a more professional approach to networking than Facebook. It gives you potential access to some powerful business identities and the ability to participate in groups with like-minded people. You'll find better opportunities to build relationships within these groups than just the home page.

Twitter

Twitter is growing in importance with 290 million monthly active users and it allows you to promote your content, increase brand awareness, and drive traffic to your website. One thing to keep in mind is that CEOs and senior executives are now spending less time on Facebook and more time on Twitter. So if they are your target market, then Twitter should be where you focus your attention. Horses for courses, as my gambling mates say.

Pinterest

Users can upload, save, sort, and manage images—known as pins—and other media content (e.g. videos and images) through collections known as pinboards. They can drive traffic to your website, but Pinterest users first have to be attracted to your pins and re-pin them to their boards. So the quality of the photo (and to some extent the information) is important. Obviously your interest in this will depend on whether there's a visual nature to your business, such as a café or dress shop.

YouTube

YouTube is a website designed for sharing video. Millions of users around the world have created accounts on the site that allow them to upload videos that anyone can watch. Every minute of every day, more than thirty-five hours of video is uploaded to YouTube. You can have your own dedicated channel, which positions you as an expert, allows you to build a community, and demonstrates your products. One huge advantage though is that YouTube lets you embed video content in your website without increasing your site's bandwidth. This means you can include video in your website without slowing down your customers' download speeds.

Instagram

Instagram has actually grown in interest for business use, particularly since Facebook changed its algorithms, which saw the cost of reaching potential customers via ads rise. Hoop Research found that the primary interest of Instagram for Australian users is their friends, with 61% of respondents indicating they follow their friends. Surprisingly though, the next biggest interest seems to be products and services, with 48% of respondents saying they like hearing about them through the platform. Comparatively, 40% are interested in photography, 39% in travel and 36% in food, which suggests that there is a real commercial opportunity for certain brands to market products and services through Instagram. It's particularly great if you have a visually focused product such as fashion.

Now that's just a brief overview of the main platforms. Others like Tumblr, VK (in Europe), Vine, and Tagged will grow and become mainstream. Actually, if we look at it from just an Australian perspective (according to Social Media News) we have:

- Facebook: 13,200,000 users
- YouTube: 12,750,000
- Tumblr: 4,900,000
- LinkedIn: 3,700,000
- Twitter: 2,500,000
- Instagram: 1,600,000

However, how you use these channels is more important than which ones you choose. The possibilities for building connections and making your name known are now virtually limitless. Your social media profile is very important in establishing your credibility. You need to tell a story that explains your work and relates to your target market. Remember, pick your ideal client and write your profile as if you were addressing them. That way it will feel to them like you are speaking directly to them and this will cause a much greater sense of engagement. Again, find someone who has done it well in that they have a great profile, and mimic their layout (or, as we call it, model their work).

Another option is to form your own community online, a collection point for like-minded individuals. There are a number of sites that help you build your own social network, such as www.ning.com and www.socialgo.com. This gives you the chance to engage with your members and not just be a digital publisher, without the need to be a code developer as the framework is provided for you as an easy-to-use platform. Prices for the base level are very low, although you can add in various features at a nominal cost. These community membership sites are different to a Facebook page as you have more control over the branding, the layout, and the adverts that run on the page. This gives you more control and looks more professional (and of course you can still network and promote it via Facebook to ensure you reach a mass audience).

Social media and broadband are not just connectors, they allow us to communicate, interact, and influence others (collectively, not just individually). To do this you have to comment on other people's posts, initiate your own posts, and also be active amongst relevant groups. This is where you will secure prospective clients, not just on the news feeds.

Become a Person of Influence

Whenever you're networking in person, or even if you're just in discussion with someone who expresses a need for a particular product or service, it helps position you as a person of influence if you have a totally reliable person you can refer them to. Now thanks to technology there's a very professional way to do that, by using an app called Referron. It's a mobile application that provides an easy process for this referral process. With just a couple of clicks on the app you can introduce both parties from your address book with an email to each explaining their needs.

Referron is simple to use, creates a very professional impression on both parties, and positions you as a person of influence. It makes sure that potential leads are not lost or forgotten, thanks to a single connection, whilst at the same time tracking your referrals. It helps you make use of your connections by providing them with quality leads and thus further builds your network.

Mark Anastasi raised a valid point when he said:

> In modern times, hard work doesn't necessarily get you anywhere. Smart work and connections do. And by this I don't mean that chasing after a college degree is the answer. There are countless college graduates hunting desperately for any job they can get their hands on as we speak. What I mean by smart work is being aware of the changes taking place in this world and working around them in an intelligent manner to make an adequate income for you and your family. And by connections I mean that networking is vital to success.

Another guru, Robert Kiyosaki, has a famous quote in this regard. "The richest people in the world look for and build networks, everyone else looks for work," he said.

There is of course a lot more to being a great networker than those attributes we mentioned at the start of this chapter. (There are actually fourteen key steps that I refer to in my training programs). One crucial component skilled salespeople will have is a strategic approach to specific targets that can help their business. A specific target is not just someone who could become a potential client, but an alliance that could steer lots of clients their way, or provide benefits in another manner. Thus a structured approach should be adopted to identify and work with companies who could benefit you.

Now, this is even easier in the online world, using two strategic avenues called joint ventures and affiliates.

Joint Ventures

A joint venture is when you partner with someone else for mutual benefit, often just on a single business transaction. This is where you start to think big. Do your research and ascertain which businesses have large databases of clients that are in your target market.

Let's look at some hypothetical examples of joint ventures. Joanne Verikios is a successful entrepreneur selling USANA health supplements. She has built a solid list of clients and followers for her blog, where she explains the health benefits behind using the best supplements in the market. Joanne builds a relationship with one of the bikini queens mentioned earlier promoting their fitness program, and they agree to a cross-promotion to each other's list. They each write a brief synopsis of the benefits of what the other is doing, giving exposure to thousands of believers who are heavily influenced by the recommendations of the gurus they follow. Thus their relationship is mutually beneficial.

Another example could involve a business coach who has a training program helping small business owners develop action plans on how to be a strategic networker. He could approach the local Chamber of Commerce and convince them of the quality of the material and the benefits for their members. They could agree to promote the training programs to their members with some personalisation added, via email, eNewsletters, and social media, and in return receive a commission of, say, 30% on all sales that result from their introduction. That way the Chamber are providing a valuable service and can earn some good income at the same time, whilst the business coach is getting exposure to a large audience that he never would have been able to tap into previously.

Of course, for real scale you could attempt to persuade a large corporation to embark on a joint venture with you. Remember earlier we spoke about the Big Data collection methods of some large corporations? Well, why not partner with a bank or an insurance company that has access to this information and thus can enable you to target selected clients much more effectively?

You can of course identify and approach potential joint venture partners yourself, and I spend quite a bit of time with my clients perfecting their approach, as first impressions make a huge difference. Another option if you need help, particularly with overseas connections, are a number of joint venture forums such as www.jvbroker.com and www.jv-network.com. You can also use www.safe-swaps.com to do an ad swap with someone else who has a similar list in terms of their size and interest, to expand your potential list. Alternatively, just approach someone you know of that has a client base that would be great exposure for you and likewise they can benefit from advertising their services to your clients.

Affiliate Partners

The other strategy we mentioned was affiliate partners. This is a relationship that involves promoting products or services solely in exchange for commission on sales. So you find someone willing to promote your product, either to their client base or just on their website, track any business that comes from them, and pay them an agreed commission for introducing that business to you. Certainly you're giving away potential profits, but these are usually clients you wouldn't have reached anyway, so think of it as a bonus. They become your external sales force, but just on a commission-only basis.

Famous American business magnate, John D. Rockefeller, would have agreed. "I would rather earn 1% off 100 people's efforts, than 100% off my own efforts," he said. And in this case you earn a hell of a lot more than 1%. Even if you give up to 50% away in commission, you're still doing well (considering these are digital products with no extra production costs) and your partners are incentivised to make some sales for you.

Likewise, it works the other way. If you build your business, not on the intent of finding customers to buy your products, but rather about finding products for your customers, you will be more successful. Create your tribe of like-minded people and then give them access to the products they'd want. You can add scale by introducing affiliates' products, which earns you a nice commission with little work required.

To do this you would first join an affiliate network, search for products that your list would be interested in, promote them, track the sales, and earn the agreed commission. These networks include Clickbank, Share a Sale, and Clix Galore. In 2015 it is estimated that commissions via affiliate marketing will total about US$4 billion for the year, and it's on a steep growth curve.

I'm pretty sure luxury brands will start to look at this avenue (with its minimal cost structure up front), as will companies looking to expand overseas without making a big commitment. Just imagine the potential of having a sales force promoting your products to clients that already trust them and buy off them, without you having to allocate any money up front. Ker-ching (or whatever sound you can attribute to PayPal, the modern cash register)!

There are five steps to implement this system with either joint ventures or affiliates:

1. Create a win/win opportunity
2. Research and find great partners
3. Approach partners and propose the opportunity
4. Negotiate the promotional strategy and form an agreement
5. Select the right platform and automate

Obviously there's a bit of work in making sure you do every step right, but the payoffs are worth the time. So whatever your business idea, whatever products or services you are making available online, your networking proceeds as normal, except you target partners that can give you extra leverage.

Mark My Words

- Networking remains integral to your success.
- Build your network, both online and offline.
- Online opens up far more potential.
- Focus on building a relationship, not selling.
- Act like a farmer, rather than a hunter.
- Always think of reciprocity.
- The various social media platforms offer different opportunities.
- To demonstrate professionalism use an app like Referron.
- Utilise others in joint ventures.
- Distribute your products through affiliates.
- Utilise the leverage technology provides.

Chapter 11
Many Hands Make Light Work

"If you deprive yourself of outsourcing and your competitors do not, you're putting yourself out of business."

—Lee Kuan Yew
Former Prime Minister of Singapore

The ideal scenario is to have a business that is not dependent on you. Many of us have a problem with delegating, which means we feel the need to be involved at most stages of the business. Thus we find ourselves drowning in detail, instead of overseeing progress and planning the course.

The way to do this is through implementing systems. There's nothing new to this of course, as an eighteen-year-old can manage McDonalds, as everything has been systemised and the staff are well trained. Systems, standards, and procedures explain in writing the things anyone working your business should do and the way they should do them. These should be in line with your overall vision, strategy, and goals. By using systems and standards for service quality as part of your everyday operations, service is more likely to be delivered consistently and naturally.

In addition to this, in the online world we use two key processes to cut back the dependence on ourselves:

- Automation
- Outsourcing

There are several aspects of the business that can be automated, with technology continuing to add more each month. You can automate your lead generation, the sales process, AND relationship building, through using an auto-responder like Aweber.

Your auto-responder sends out a series of prepared emails to your contacts at set intervals. So you can establish a campaign whereby, say, twice a week you send out emails touching on a different facet each time. It then tracks the analytics behind the campaign and can tell you how many of your clients opened the emails, clicked on a link, or forwarded it on. Thus you also get a better understanding of what's working and what's not, so you can tweak the next campaign to ensure better results.

If you have to ship physical products (from your site, Ebay, or Amazon as we mentioned a few chapters ago) you can use the Amazon fulfilment service (called FBA). You ship your product in bulk to one of Amazon's fulfilment centres and they take care of the rest once orders are processed via your site. This means that FBA automates and takes care of:

- Shipping individual orders—saves you being bogged down running to the post office every day.
- Handling returns—so you can focus on getting more products and building your business.
- Managing feedback—Amazon takes the hit, not you as a seller, if there happens to be negative feedback related to the fulfillment experience.
- Enquiries—Questions related to order status and shipment arrival are handled by Amazon.

Tim Ferriss, in his excellent book *The 4-Hour Work Week*, opened many eyes to the potential of systemisation and automation, and the resultant benefits from removing yourself from the equation.

Now you can start to see where the lifestyle component of these online businesses kicks in, freeing you up to create great content for your products and outsourcing the rest to experts.

Outsourcing just make sense. These days we're more than comfortable doing it around the home. Previous generations were much handier and spent more time fixing problems. But these days we don't fix our own plumbing, electrical, car, or other household repairs. In fact, most people even outsource maintenance jobs like mowing, and we are comfortable using specialists in their fields.

So we apply the same process to our online venture. Much of the back end parts can be done by specialists who have a better understanding and an ability to perform the task quicker. The two main reasons to use outsourcing are that it enables you to reduce your costs, by hiring cheaper labour in developing countries like the Philippines or India, and it allows you to use your own time to focus on your core activities.

There are many places you can find outsourced specialists, for example I have used www.onlinejobs.ph and www.easyoutsource.com. I have secured some contractors who are very talented, and some whose work ethics were questionable, so you have to do some due diligence first (like you would with any remotely based employee). One aspect to consider though is that if you hire an individual and they go missing, your productivity is put on hold. Because of this, some people prefer to pay a bit more to hire a company that allocates internal outsourcers. That way if a staff member leaves, then they simply allocate someone else to your tasks.

Alternatively, if getting the maximum benefit out of the social media side is an area that concerns you, you can outsource this as well. Check out www.socialmediaworldwide.com where they use their expertise to not only design and build your pages, but also to load appropriate posts for you every day. Their standard plans start at $497 per month, but keep your eye out for specials from time to time. They'll drive the traffic to your site, allowing you to focus on developing great content for your potential clients. As you know, getting traffic (visitors to your site) is crucial if you want to grow your business, so consider using the specialists.

We've already mentioned Fiverr previously, but there are other sites also where you can find some specialist help for your project. These include www.elance.com, www.upwork.com (formerly called www.odesk.com) and www.freelancer.com. These sites operate a bit like an auction process, whereby you post details of the project you want done and a large number of freelancers will forward you their proposal and price, leaving you to choose from all the talent available. Having talented people available to hire by the hour or the duration of your project gives you flexibility and an advantage over your competitors.

If you're in the service industry then outsourcing can bring spectacular results. Imagine if a client wants something done but doesn't drop it off to you till nearly closing time. You can have someone in a different time zone do the work and the finished product can be emailed to your client before he even gets to work the next morning. Now that's service. There are downfalls of course, one of which is communication. Most of the outsourcers are in developing countries so selecting the right ones initially is crucial. If you are contemplating this, go to www.upwork.com and download their free eBook *Make it Happen*. It gives you lots of great tips on how to proceed.

You can of course hire a virtual assistant, graphic designer, or webmaster from within Australia or another developed country if that is your preference. There are some very talented operators who like to work from home utilising their skills in this manner.

If you're using outsourcers you still have to monitor their activities and measure their progress. You can ask them to report to you daily, with some simple answers to template questions such as:

1. What tasks have they completed that day?
2. What tasks are outstanding?
3. What questions do they have for you?
4. Are there any issues to raise regarding the tasks?

You could use software like Harvest for time tracking, which enables you to measure outputs.

It makes sense of course to combine automation and outsourcing. Use this external expertise to design your site and back end, implement the systems that automate certain processes, and then outsource any repetitive tasks to a virtual assistant. This enables your business to grow and frees up your time to promote and oversee. This is called scalability, as in the process of being able to increase sales transactions and thus profits, without needing significant additional resources.

In any business you want a good team around you—as we've seen already, they don't necessarily have to be employees. Build relationships with external professionals to ensure they can guide you through the processes. Accountants, lawyers, and even mastermind members are great resources, so ensure they know your business.

As a business coach and strategist I help my clients do whatever it takes to get them to the next level in business. Some require my strategic input, but many just value being held accountable and having someone to bounce ideas off. Almost like an independent director. For whatever reason, it seems to work, as they all become successful. They are more confident after my feedback and more committed to action. So build your external team and work with them.

You should also have an exit strategy in mind before you start any business. Do you want to sell it at a certain stage down the track? If so, then scalability makes it an attractive proposition for a potential buyer, and having systems in place is imperative. In many cases this is where the real money is made.

Now when you're working on your action plan, make sure you keep these concepts in mind. What systems can you develop, what and where can you outsource, what processes can you automate?

Mark My Words

- Establish processes not dependent on you.
- Introduce automation wherever possible.
- Read *The 4-Hour Work Week.*
- If you're selling products use Fulfilment by Amazon.
- Outsource to industry specialists.
- Focus on the parts you do best.
- Build your team of external support resources.
- Ensure your business is scalable.

Chapter 12
Action Planning

"Have a bias towards action—let's see something happen now. You can break that big plan into small steps and take the first step right away."

—Indira Gandhi

It's great to have ideas, but the only way you'll get results is through action. How many times have you seen something and thought, "That's a good idea," or, "I could use that in my business"? It can be a video you watched, something you heard at a seminar, or just a topic someone mentioned in conversation. It happens all the time, right? Many of us are exposed to the same information, but successful people (and now you as a Game Changer) will look to action the things you've learnt. That doesn't mean that you have to implement them straight away, as it might not be appropriate in your current circumstances, however you record the details in a dedicated place so they're always available when needed. So set up an ideas book today and keep it handy. Of course, if you can use that idea right now to improve your business, then do it. Don't procrastinate. Follow the processes within this book and just do it.

By now you'll know that if it's going to happen, you have to take the initiative for yourself. No-one is going to give it to you (well, except for the great content provided here). No-one should have to hold your hand and give you a kick start. It's up to you. You can transform your thinking and take the actions to control your destiny. To do this you have to be taking action every day. Your thoughts should be focused on opportunities, even from moment to moment. Get clear and then that reticular activating system will kick into gear. Success will follow as you allow your creativity to expand your frontiers.

You'll find it's like anything in life. Challenging at first, but then you get the hang of it and everything seems to snowball. Momentum creates momentum. To do this though, you have to use the fire inside you, use it to push you on, use it to cultivate the relationships you need to be successful. People gravitate to those who are confident and do things with a passion. Bring out that side of you. Don't settle for second best, or waste your time doing things you don't want to. Learn to say no, so as not to overstretch yourself when you need to be focusing.

As a sports fan I know you're not going to affect anything, or make a difference, by sitting on the sideline. Getting in the game is the first step, then take it to the next level by being a Game Changer.

Okay, it's time to take those first steps. What changes to your daily behaviours will you introduce right now? What steps do you need to do to lay the foundation for the success you want to be?

Set your goals out like stepping stones. Mind-map the process. Small steps, different small steps, then something a bit chunkier. Hopefully you'll be planning a better morning routine with exercise, healthy eating, and something creative to get you thinking in the right vein. Introduce your routine and stick with it, and never forget we undermine our own reality when we settle for mediocrity. Be the best you can be. It's time you took back control.

Now let's focus on the business opportunities. Remember, performance is a function of action (in that there's a consequence for action and for inaction). As Pablo Picasso said, "Action is the foundational key to all success." You're setting up the dominoes and getting ready to knock them down and create that momentum. Let's get a plan in place, or at least an idea that germinates in your head whilst you do some further research around it.

Everyone needs a plan. It's a roadmap to follow. But it's not the plan itself that's important—it's the process you take to get there. You may not even need to do a detailed business plan, and certainly never hire someone to do it for you. That must sound strange—a business coach with an MBA, a strategist with a love of following a process, telling you that you don't need a detailed plan. That's a fall back to my early career days in the bank analysing large corporate lending deals, where you could always tell how much input the business owner had into the document and thus how often he would actually read it. You don't want a big, thick document that's going to sit in the bottom drawer and only be used to convince a lender to give you some finance. You certainly don't need to do something like three years of cash-flow projection—things change too quickly these days for that. You want a document that will act as your roadmap.

You need to understand the basics and get your concept down on *one page*. In fact, that's a process I take all my personal clients through, getting clear focus on a one-page plan. This can cover off topics like:

- Your vision and/or mission statement
- Your business strategy and channels (e.g. online distribution of third party products)
- Your competitive advantages
- Your strategic initiatives (projects to drive sales)
- Likely expenses and when they'll be incurred
- Realistic revenue expected
- Key partners
- Your exit strategy

This is the new age, time for doing, not time for detailed planning. Read books or ask for help with the skills you need to develop, watch videos to learn how to do certain things, and listen to podcasts to help stimulate your ideas. There are some great podcasts that can play in the background on your computer, or on your iPod whilst you're doing something relaxing (not that you'd waste much time on your personal Facebook page these days). Examples of these include www.tropicalmba.com and www.buildandlaunch.net, as well as the more traditional ones you'll find on iTunes.

Better still, you will soon spend your travelling time in the car listening to podcasts that will not only inform you, but they'll keep you inspired. You might work shorter hours, but you'll use some of the time you've saved to help you stay relevant and creative through the thoughts they provide.

In time, plan on having your own podcast to see your influence explode. They are incredibly rich sources of information compared to written stories, not just in terms of the reported facts but in terms of imparting tone of voice and subconscious patterns of social interaction that frame an opinion or argument in a much more specific way. Thus through podcasts, like videos, you build a stronger relationship with the listener.

Get a concept, model someone who has done something similar, listen to the experts, implement it cheaply, and then improve it along the way.

One of the leading exponents of social media marketing is Corinna Essa. I asked her what the online experts were doing to drive traffic to their sites and she confirmed they mainly use:

- Joint venture partners
- Social media
- Paid advertising (including Twitter ads, Facebook ads, LinkedIn ads, Google AdWords)

However, she stated, "They also use Search Engine Optimisation, email marketing, ad swaps, buying email lists, YouTube, YouTube ads, text marketing and banner ads." So that's a pretty comprehensive list, but remember you build your efforts on the various platforms over time. In completing your action plan you can stage the process over many months to ensure you master each aspect before moving on.

I also asked Corinna if any of the social media platforms are more suited to particular types of businesses. She replied, "Not necessarily with Facebook and Twitter, as most people and types of businesses use them, however with LinkedIn, it's more for corporate companies and people looking for jobs or business partnerships, as it attracts professionals. With Pinterest, it's more suited for arts and crafts companies, as well as clothing and food companies." Again, target those platforms that you feel will enable you to reach your target market.

Implement It

Do what successful people do—they take the next step. Richard Branson was already very successful thanks to Virgin Records, when he had an idea about improving the quality of air travel and injecting his brand of fun into the process. The same day he thought of the idea, he took the next step. He phoned Boeing to see how much it would cost him to lease an airplane and was able to do a good deal that allowed him to return the plane if it didn't work out. That gave him the comfort and the initiative to push forward. Once you have your idea in mind, take the next step.

If you believe in yourself and you like to win, then play to win. Yes it's a game, you're in competition, and to survive you have to keep winning.

I'm sure you've also heard the often-quoted statistic that only 5% of people have their goals written down and those 5% make more money than the other 95% combined. The act of writing and referring to it daily helps keep you focused and on track. It ensures you are thinking and acting in the manner you need to if you're going to achieve those goals. And if that's what successful people do, then it's a no-brainer for you to do it too.

Here's another great thought. You can use numerous channels to reach people who have a preference as to how they spend their time or receive information. Take this book for example; the contents can easily be re-purposed into many formats, including:

- Podcasts
- Seminar workbooks
- Keynote presentations

- Video series
- Webinars

This approach enables you to create at least five pieces of content from your original concept, resulting in a larger audience than if you just restrict yourself to one dimension. So by taking action in one area you open up other possibilities.

Unapplied knowledge is useless. You might remember the old saying 'Knowledge is power'—well, that doesn't cut it these days. *Application* of knowledge is what's needed. Apply that knowledge in a way that uses the leverage of technology to engage mass markets. That's power. That's a Game Changer.

Mark My Words

- This time provides the best opportunity you'll get.
- Don't procrastinate, get into the game now.
- Model the experts to have your shortcut to success.
- Performance is a function of action.
- Introduce your routine.
- Write your goals down.
- Select the platforms of interest.
- Complete your one-page action plan.
- Play to win.
- Apply your knowledge.
- You have the advantage of fast, easy implementation.

Chapter 13
Now it's Up to You

"I can give you a six word formula for success—think things through, then follow through."

—Eddie Rickenbacker
World War I fighter pilot

"I love that one, but here's my four word formula for success—Become a Game Changer."

—Mark Brigden

Imagine having a business that continues to function whilst you're travelling the world. You could be enjoying the delights of Sydney, Hong Kong, Zurich, or London, or even surfing in Hawaii; it doesn't matter as you're still connected to your business. Or, reflect on how good it would feel to finally publish that book that gives your life meaning. Maybe it's the respect you get when you become more influential and regarded as a person of influence, or is it just the feeling of satisfaction that you have embraced technology and feel relevant in a changing world?

Whatever it is that drives you, this is the perfect time to make it happen and to reap the benefits. You can accomplish so much without having to part with a lot of money. That's the beauty of where technology is at now. Your office can now be your laptop and your customer base can now be worldwide. Your files are stored on Dropbox—so accessible wherever you are. It just needs the commitment to take that first step.

You've read this far, so you must have a desire to do something, to make a difference, to have a go. Congratulations, this desire will help drive your success. Most people, of course, do not act on their desires, they let life control them and then complain about the awful hand they've been dealt. They just don't want it as much as you, and that's why you'll be successful. Let this desire inflame your passion to take action.

At the start of this book we spoke about age not being a deterrent; in fact you should regard your experience as a benefit. But let's look at the other end of the scale, can someone young do something to take advantage of technology? Well, *Entrepreneur* magazine highlighted a perfect example right here in Australia. In their story they used data compiled by Outrigger Media, who assessed the top-earning channel on YouTube in the one category of food and cooking.

There is a channel called Charli's Crafty Kitchen, which features videos of a delightful eight-year-old Australian, named Charli, sharing baking tips. Charli began making her videos in 2012, when she was just six years old. Her five-year-old sister, Ashlee, also plays a role as chief taste tester. According to Outrigger's estimates, Charli's channel generates an average of $127,777 in advertising revenue per month. That's taking into account YouTube's cut of the revenue. The channel gets an average of twenty-nine million views each month—in March 2015, it had 29,133,270. Over $120,000 a month. Can you imagine making that much money at eight years old? Can you imagine making that much money at your current age? Well, you can now. If an eight-year-old can do it, surely you can have a go. She had the courage to do it, and obviously the ability to just be her cute self.

No fear, just have a go—isn't that the Australian way?

I think it's fitting that we reflect on the words of Tim Berners-Lee, the inventor of the World Wide Web, as he put it prior to the turn of the century:

> The next century is going to turn our world upside down. The Internet combines people and ideas faster than they have ever seen combined before. And that combination changes everything. The basic social conventions of the industrial era—the stable career, the 9 to 5 job, the gradually (but steadily) increasing salary—were all built

> around the notion that people moved their bodies in response to information.
> If you wanted to buy something you went to a store. If you wanted to build something you worked in a factory. In the net economy, the creation of value doesn't require that kind of physical movement. Income accumulates not in the form of cash but in the form of clicks.
> The great thing about technology is that it forces us to figure out the world from scratch. In so doing, it gives us a chance to rediscover what's really important. So maybe the 21st century won't turn your world upside down. Maybe it will turn that world right side up.

As you've seen highlighted throughout this book, the critical factors you will need to develop (or display) to be successful in the new world of business include:

- Inspiration
- Creativity
- Flexibility
- Innovation

I have no doubt you have the ability to do this, however don't pass up this opportunity. Act before it's too late.

The speed of Internet connectivity is about to go to another level, with 4G and 5G quicker than infrastructure, so take advantage of the major shift that has been happening and will continue. A study in early 2015 commissioned by Westpac Bank found two thirds (sixty-seven per cent) of Australians report having had ideas about new businesses, products, or services in their lifetime, with an average of 10.9 ideas each. But more than half (fifty-four per cent) have forgotten ideas, at an average of 7.3 each. This could amount to around sixty-four million forgotten ideas nationwide, according to Westpac's Business Ideas Report.

The survey of 1000 Australians and 1000 small businesses found that while two thirds of people have had ideas about new products and services, around one in five (twenty-two per cent) have acted to bring a new idea to market. Don't join those statistics and regret it. It's the things we don't do, which we should have done, that haunt us in later life. Luckily there's been lots of publicity around these types of issues in social media recently, which has raised awareness of this need for fulfillment.

In fact another study, this time by Bankwest in 2013, found that Australian women aged over sixty-five are starting their own businesses at a rate unmatched by any other generation in the last ten years. Is it any coincidence that they have been the ones also jumping on to social media? They have seen the potential and they want to do something meaningful. Now they can, and so can you.

Do you know what a Game Changer would do now that they've finished this book? That's right, they'd go back to the start and look at it from a different point of view now that they have been exposed to the complete message. They'd look for ideas, processes, angles that suit their interests. They'd look for inspiration as to the right format for their concept, they'd get their mind around the questions to ask as part of their market research, and they'd commit to making it happen.

It's a long road, but it's not hard work if you're having fun doing it.

Now it's up to you ...

BECOME A

GAME CHANGER